T0251030

Cell Physiology and Genetics of Higher Plants

Volume I

Author

A. Rashid
Reader (Associate Professor)
Department of Botany
University of Delhi
Delhi, India

CRC Press
Taylor & Francis Group
Boca Raton London New York

CRC Press is an imprint of the
Taylor & Francis Group, an informa business

First published 1988 by CRC Press
Taylor & Francis Group
6000 Broken Sound Parkway NW, Suite 300
Boca Raton, FL 33487-2742

Reissued 2018 by CRC Press

Library of Congress Cataloging-in-Publication Data

Rashid, A. (Abdur)
    Cell physiology and genetics of higher plants.

    Includes bibliographies and index.
    1. Plant cells and tissues. 2. Plant cytogenetics.
I. Title.
QK725.R33 1988   582'.087   87-9349
ISBN 0-8493-6051-X (set)
ISBN 0-8493-6062-5 (v. 1)
ISBN 0-8493-6063-3 (v. 2)

A Library of Congress record exists under LC control number: 87009349

ISBN 13: 978-1-315-89138-5 (hbk)
ISBN 13: 978-1-351-07048-5 (ebk)

Visit the Taylor & Francis Web site at http://www.taylorandfrancis.com and the
CRC Press Web site at http://www.crcpress.com

PREFACE

In the beginning of this century, Haberlandt performed his pioneering experiments on the culture of plant cells. Simultaneously he had visualized the potential applications of this technique, and thus was born plant cell biotechnology. In the intervening period, the technique of plant cell culture has been perfected, but plant cell biotechnology remains in its infancy. The reason for the slow progress is inadequate information about physiology and genetics of higher plant cells. In this series a synthesis of the concepts of cell physiology and genetics, the basic disciplines for cell biotechnology and genetic engineering, is presented. A broad and informative view of these disciplines will be appealing to plant physiologists, geneticists, cell biologists, plant breeders, and biotechnologists.

Volume I comprises four chapters. Initial chapters devoted to fundamental aspects (Cell Multiplication and Cell Differentiation) concerning cell physiology are basic to the theme of biotechnology of higher plants. Included in the chapter on Cell Differentiation is an account of biosynthetic potential of higher plant cells. The next chapter, Cell Totipotency, highlights the regeneration potential of higher plant cells, for micropropagation of plants. Following this is the chapter entitled Induction of Haploid Plant/Cell. Developments in this field have led to the beginning of a new era in crop improvement.

Volume II comprises five chapters. Of these, four chapters are devoted to cell genetics: Protoplast — Isolation and Cell Regeneration, Cell Modification, Cell Fusion, and Cell Transformation. These chapters concern genetic engineering of plants. The final chapter, Cell Preservation, for germplasm storage, concludes the series.

This work is an outcome of the freedom a teacher of this department enjoys in framing a course and offering it to students in lieu of an elective paper for an M.Sc. Degree in Botany. I acknowledge the indirect contribution of students for stimulating discussions. Two of my former research students, now teachers of botany, Dr. Shashi Tyagi (nee Bharal) of Gargi College and Dr. Paramjeet Khurana (nee Gharyal) of Khalsa College, were requested to go through these chapters at the first draft stage and offer critical comments. I am appreciative of the help received from them. Thanks are also due to my research students, Mr. M. A. Mallick, Miss Rajni Dua, and Miss Manju Talwar, for their help in completing literature surveys.

I would also like to acknowledge the much needed encouragement and invaluable help received from my wife, Dr. Zakia Anjum, in undertaking and completing this work.

A. Rashid

THE AUTHOR

A. Rashid, Ph.D., is Reader in Botany at the University of Delhi. Dr. Rashid received his degree from the University of Delhi in 1968 and since then he has been a member of the faculty. Earlier, his research interests were restricted to developmental physiology of lower plants. Later, in 1972, he was attracted towards higher plant cell physiology and completed a postdoctoral fellowship in the laboratory of Professor H. E. Street at the University of Leichester, U.K. From 1980 to 1982, he was also awarded the Alexander von Humboldt Fellowship to collaborate with Professor Dr. J. Reinert, at the Freie Universität Berlin for the induction of haploids in culture of isolated pollen grains.

Dr. Rashid has published 60 research papers and a few review papers. His primary interest is cell physiology, employing pollen and protoplast as systems.

CELL PHYSIOLOGY AND GENETICS OF HIGHER PLANTS

Volume I
Cell Multiplication
Cell Differentiation
Cell Totipotency
Induction of Haploid Plant/Cell

Volume II
Protoplast — Isolation and Cell Regeneration
Cell Modification
Cell Fusion
Cell Transformation
Cell Preservation

TABLE OF CONTENTS

Chapter 3
Cell Totipotency..67

Chapter 1

CELL MULTIPLICATION

I. INTRODUCTION

Plants contribute to a wide range of industries including those of food and feed, fabric and furnishing, chemical, paper, and energy. Some of these industries are central to our needs and others are peripheral. Due to an ever increasing population and the consequent decrease in land area for cultivation and forestry and dwindling energy resources it is imperative for plant scientists to have continuous plant improvement programs. In the past this has been accomplished through plant breeding. However, the potentials and properties of plant cells remain to be explored. Plant cells in culture, analogous to microbes, are seen as a system for translating the conceptual advances of molecular biology to plant biology and ultimately to human welfare. This introductory chapter on cell multiplication is basic to the theme of higher plant cell biotechnology.

For an understanding of cell multiplication, a peep into the past will be in order because the history of plant cell culture is quite instructive.

II. HISTORY

In the beginning of this century a German botanist, Haberlandt,[60] for the first time visualized that the culture of isolated tissue or cells from higher plants would provide an insight into the properties and potentialities of a cell, the elementary unit of the whole plant. As for potentialities, he pointed out the possibility of obtaining embryos and then whole plants from vegetative cells. In fact, his postulation turned out to be true when another German botanist, Reinert,[117] reported the induction of embryos from the cells of carrot. Since then it has been possible to obtain embryos from cells of many plants.

Haberlandt attempted to culture palisade cells of *Lamium* and *Eichhornia,* epidermal cells of *Ornithogalun,* and epidermal hairs of *Plumonaria* on Knop's mineral medium supplemented with sucrose. These cells remained alive for some time, but failed to divide. The failure was due in part to selection of cells. Cells from palisade tissue, epidermis, and epidermal hairs are fully differentiated and relatively refractory to division. Haberlandt also did not supplement the medium with growth regulators, necessary for cell division, but not known at that time.

Although Haberlandt was unsuccessful in the culture of higher plant cells, he made three remarks which are the reflections of his foresight. One, the differentiated cells cease to divide. This was not because these cells lose the potential for division, but the stimulus for division arises either from the whole plant or from a part of it. He even pointed out that the growth stimulus might be "growth enzyme". Second, it would be worthwhile to culture cells along with germinating pollen grains. This suggestion was, in fact, based on the observation of his contemporary, Winkler,[170] who found that on pollination, the pollen tube stimulated the growth of ovary and ovule. Third, one could add to the nutrient medium an extract from vegetative apices. Alas! Haberlandt would have done it, because now it is known that cells from a meristematic tissue can undergo a few cycles of division without an external supply of hormone. One such example is that of cambium cells. In fact, cambium cells[48] of willow and poplar proliferated for some time on a simple mineral-sucrose medium. This is the first instance of cell culture of higher plants, by a French botanist, Gautheret.

Soon after Haberlandt's work, nearly mature embryos[61] from a number of crucifers were reared to plantlet stage on a mineral medium supplemented with sucrose. This was the beginning of in vitro studies on reproductive biology of higher plants. This aspect is beyond the scope of this book.

Since Haberlandt's pioneering attempt and the success of Gautheret in initiating cultures of higher plant cells, there was no significant progress in the continuous culture of higher plant cells until 1934 when an American botanist, White,[163] succeeded in initiating culture of tomato root on a mineral medium supplemented with sucrose and yeast extract. The lateral roots, developed in culture, on isolation from the main root served as explants for subculture. The nutrient formulation employed for culture of root is described after the pioneer, as White's medium. Later, yeast extract was replaced by three vitamins: pyridoxine, thiamine, and nicotinamide.[164] Culture of root, in a strict sense, is not a culture of tissue or cell, it is instead an organ culture.

Continuous proliferation of cells was reported in 1939 when in a rare coincidence three independent workers — Gautheret,[49] Nobecourt,[107] and White[165] — succeeded in establishing continuous culture of cells in the form of an amorphous mass, described as callus. Gautheret as well as Nobecourt, both from France, successfully cultured cambium cells from carrot root on a mineral medium supplemented with auxin, IAA. The success is to be ascribed to inclusion of auxin, whereas White, from America, achieved the growth of tumor cells of tobacco hybrid (*Nicotiana glauca* × *N. langsdorffii*) on a simple mineral medium without hormones. This was possible due to the fact that tumorous cells are autonomous in their hormone requirement for division.

Thus, the works of White, Gautheret, and Nobecourt are the pioneering attempts at culture of higher plant cells. Since then a great deal has been discovered about induction of cell proliferation.

III. INDUCTION OF CELL PROLIFERATION

Proliferation of cells has been possible from almost every part of the plant, vegetative as well as reproductive. The multiplication of cells results into a new type of growth, an amorphous mass of cells, the callus. Callusing has been possible from cells derived from root, stem, leaf, flower, embryo, endosperm, and even pollen. For initiation of a callus, generally a differentiated tissue, either an entire one or part of it serves as an explant, for culture, on an induction medium jelled with agar. Callusing is also possible on a shallow liquid medium, but agar surface is preferred. In a differentiated tissue, cells are not in a state of division. Therefore, there is a transition period from quiescent to active state.

A. Transition from Quiescent to Active State

This period, duration of which depends on physiological state of cells in the explant and the chemical conditions employed for induction of cell proliferation, is divisible into different phases.[1] To begin with, there is an induction phase during which the metabolism is activated and cells prepare to divide. This is followed by phases of active synthesis and active proliferation of cells.

Parenchyma cells from storage tissue of *Helianthus tuberosus* (Jerusalem artichoke) have served as a suitable system for the study of different phases of this transition. On an induction medium, comprised of mineral salts and an auxin, 2,4-D, the explants from this tissue show rapid proliferation of cells (Figure 1) which is synchronous to begin with. This suggests that cells in this tissue have been brought to a halt at a similar stage of development. It also supports the viewpoint that cells in a differentiated tissue retain the capacity for division but are repressed. However, nothing is known about the nature of the repressor in this or any other tissue. Also, it is not known why cells in the process of differentiation lose their capacity for division. Numerous changes have been described in a tissue prior to proliferation.

Initial changes in a tissue in response to excision, within minutes, are the appearance of

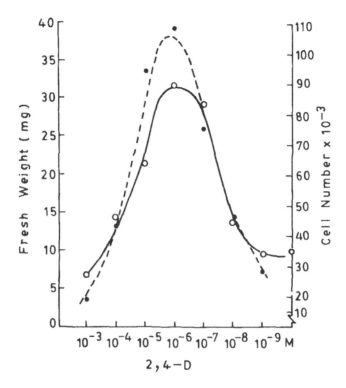

FIGURE 1.   Increase in cell number and fresh weight gain of tuber tissue of Jerusalem artichoke as a function of 2,4-D. The tissue was grown for 7 days on medium containing minerals, sucrose, and auxin. (From Aitchison, P. A., Macleod, A. J., and Yeoman, M. M., in *Plant Tissue and Cell Culture*, Street, H. E., Ed., Blackwell, Oxford, 1977, 267. With permission.)

polysomes[6] and a rapid increase in the rate of gaseous exchange.[171] These are more pronounced when the tissue from tuber is maintained on a callus-induction medium.

Possible alteration in cell permeability before the onset of dedifferentiation is inferred from membrane modifications seen during the first hours of culture of leaf explants of *Datura innoxia*.[21] Pinocytosis, loosening, outgrowths, and infolds of plasmalemma and plastid membrane invagination were seen in sucrose-containing medium. Kinetin and, to a lesser degree, auxin magnified these responses.

Enhanced respiration in an excised tissue taken from tuber of Jerusalem artichoke suggested the possible involvement of mitochondria. In particular, the activation of dormant tissue results in an early synthesis of DNA which reaches a maximum at 3 hr, much before the beginning of S phase (12 hr). By electron microscopic autoradiography[44] it was possible to determine that plastids and mitochondria were the organelles responsible for this early synthesis while label appeared after 15 hr in DNA of the nucleus.

Prior to proliferation, there is an accumulation of RNA in tuber cells, and it is a stepwise process. The first increase of 30 to 40% occurs during the first 6 hr and this is possible with or without auxin in the medium. Therefore, it is very likely that this is in response to excision alone. The second increase, about 120%, is simultaneous with DNA replication and the last, 300%, occurs just before cell division. The synchronous divisions in tuber tissue of *Helianthus* in response to auxin 2,4-D have been correlated with increased levels of chromatin-bound RNA polymerase and an increase in the rate and degree of processing of rRNA.[90]

As for changes in different enzymes, a measurement of the activities of glycolytic, mitochondrial, microbody, and hydrolytic enzymes during cell dedifferentiation of maize tissue, induction and establisment of callus showed that it was comparable to that of cells present in the plant. This led to the inference that there is no rearrangement in enzyme pattern during callus formation.[175] However, there are marked differences in protein profiles of quiescent cells and cells about to divide.[1] In particular, nonhistone proteins (NHP) undergo significant changes during cell proliferation which imply their role in this process.[55] In tobacco pith tissue, prior to proliferation, low-molecular-weight NHP bands, which were intense in pith tissue, decreased when callus was forming, and high-molecular-weight NHP bands increased drastically which decreased when cell division declined.

In view of these varied changes in a tissue, prior to proliferation, it is quite natural to assume that it is a function of a number of factors.

B. Factors Regulating Cell Proliferation

Of the factors regulating initiation and continuous proliferation of cells the important ones are given below.

1. Wound Reaction

Excision or wound reaction is an important factor in the induction of cell proliferation. Wounding alone is sufficient to induce proliferation at the cut surface in many dicots and at times may result in the formation of a substantial amount of tissue. By contrast, in many monocots cell proliferation usually does not occur in response to wounding. Therefore, an understanding of wound reaction is likely to reveal the regulation of cell proliferation.

The importance of the substance(s) liberated at the cut surface in response to wounding, wound hormone, has been emphasized time and again in plant physiology literature. Cell division, however, is not the function of wound hormone alone, but is a result of an interaction between wound hormone and hormone(s) present in the tissue, the leptohormone.[60] Formation of a substantial amount of callus at the cut surface of meristematic tissues, such as cambium,[48] can be explained as a function of wound hormone and endogenous substance(s) present in the tissue at the time of excision.

In brief, wound reaction is the first trigger to proliferation. However, excision or isolation is a physical process and cell proliferation in response to wounding, a biological adaptation of living tissue, in the final analysis is an induced chemical reaction.

2. Chemical Environment

More important for initiating cell proliferation is the chemical environment provided to tissue. Even in an intact tissue (leaf, seedling, or an embryo) callusing is possible in response to proper chemical environment. In the absence of wound reaction, chemical environment serves both for induction of cell proliferation as well as for cell multiplication. Otherwise, chemical environment is essential for cell multiplication, the growth response.

Higher plant cells, unlike lower plant cells, on isolation are incapable of division without an external supply of hormones. Again, a distinction is desirable between initiation of cell division and the continued cell division. For the first response, cell proliferation, one has also to account for an interaction between the external supply and the endogenous level of hormones. The endogenous level is variable from tissue to tissue and even in the same tissue, it depends upon the time when the tissue is excised. Cells from a meristematic tissue, due to endogenous level of hormones, are capable of undergoing a few cycles of division without an external supply of hormones. Cells of cambium is one such example. Another common example is juice vesicles from lemon fruits. The juice vesicles of a mature lemon fruit[78] are no longer mitotically active because they do not manifest mitotic activity in response to excision injury alone. However, the ability of excised juice vesicles to manifest indefinite cell proliferation[79] in the form of a callus tissue on a simple mineral-sucrose medium, devoid of growth substances, indicates that the capacity for division is not irreversibly altered in this tissue in vivo, but has been arrested by an unknown biological factor.

The role of an endogenous factor(s) in regulating initiation of cell division becomes clear from a work on induction of cell division in the epidermal cells of *Torenia fournieri*[26] on isolation from stem segments. Divisions in epidermal cells stopped immediately when the epidermis was separated from subjacent tissues. However, division in isolated epidermal cells could be induced by adding asparagine, alanine, and glutamine to the medium. Growth substances acted synergistically with amino compounds. Therefore, amino compounds are possibly one of the endogenous factor(s).

Of the hormones, auxin is an essential requirement for cell proliferation as well as cell multiplication. A cytokinin is beneficial, but not essential. However, for some tissues, such as tobacco pith, inclusion of a cytokinin is especially beneficial. Therefore, an increase in growth of tobacco pith tissue in response to external supply of a cytokinin along with an auxin is a standard bioassay for cytokinin. Gibberellins, either in the presence of auxin and cytokinin or alone, do not have any specific role and are rather inhibitory. However, exceptionally, for the induction of proliferation from mature and dried endosperm tissue, some factor(s) contributed by germinating embryo is essential,[23] and gibberellins replace this "embryo factor" essential for proliferation of mature endosperm.[135]

As for auxins, IAA, a naturally occurring auxin, is readily broken down by cells. Therefore, a synthetic auxin, 2,4-D or NAA, is preferred for inclusion into the medium. The concentration required is in the range of 10^{-6} to 10^{-5} M (about 0.2 to 2.0 mg/ℓ) for an auxin, and between 10^{-7} to 10^{-6} M (about 0.02 to 0.2 mg/ℓ) for a cytokinin. Exceptionally, for some tissues a high level of auxin is required. Beetroot[62] is one such tissue where 10 to 25 mg/ℓ of IAA or NAA is required and cotton[73] is another example where up to 50 mg/ℓ of NAA is required for induction and growth of callus. Also, for the endosperm of *Cocos nucifera*,[80] in addition to embryo factor, a high concentration of auxin 2,4-D (50 mg/ℓ) is required. Of the auxins, 2,4-D tends to repress organogenesis from callus tissues of dicot plants; therefore, NAA or IBA is preferred, but 2,4-D seems to be an auxin of preference by the monocot tissues.

The variable response of different tissues to different growth regulators and their concentrations calls for an understanding of physiological status of explant.

3. Physiological Status of Explant

An important factor in cell proliferation is the physiological status of the explant. This explains the variable responses of tissues isolated at different times of the year and even at the same time when the explants are from different parts of a plant.

When tissues are taken in spring there is increased frequency of proliferation. This is particularly true of woody plants such as *Populus*.[18] The variable response of stem explants of tobacco has been recorded in many laboratories, particularly when the explants are from different levels of plant. This is explained in terms of endogenous levels of hormones which change in a plant[134] with a distance from the apex. In storage tissue of Jerusalem artichoke this is seen in terms of a variation in time taken for proliferation. When tissues are taken during the first 5 months of storage of the tuber, proliferation is possible within 20 to 24 hr after excision. This lag phase increases gradually, with the period of storage, reaching to 60 hr if the tissue is stored for 1 year. The lengthening of lag phase for proliferation was found to be associated with a change in the metabolic status of the tissue; in particular, the amount of RNA declined with a period of storage.[86]

In a fruit, ripening is described to be accompanied by a loss of cellular functions. Explants from fruits of apple and pear showed differential capacity to enter cell division. Tissue from apple cortex was able to proliferate, in response to auxin and cytokinin, long after maximum climacteric, whereas pear cells lost their capacity to divide very early.[83]

Physiological status of explants, in a strict sense, is an expression of endogenous level of hormone(s) present in the plant. Therefore, in such a consideration, along with promoters, the role of inhibitors should not be ruled out. A good example is *Gossypium barbadense*.[73] In order to induce cell proliferation and continued callus multiplication from cotton seedlings along with auxin, an antioxidant, dithiotheritol, and an auxin protector, ascorbic acid, were beneficial.

4. Genotype and Cytological Status of Explant

Different genotypes are variable in their response to proliferation. This is particularly true of cereals where some cultivars readily respond to form a callus and others fail to do so. In addition to the all-or-none response, different tissues are variable in their requirement for auxin. The variety "Violet de Rennes" of Jerusalem artichoke is auxin-requiring whereas variety "Blanck Sutton" readily proliferates in absence of an auxin.[46]

As for the cytological status of an explant, it is generally believed that a tissue consists of cells which are fairly homogenous, but this is rarely so, particularly when tissues are of stem pith or root origin. Also, it is a common observation that not all cells in an explant proliferate to form a callus and of those that do so, not all multiply at the same time. There is some sort of selection; only certain cell types proliferate in a particular chemical environment. This was clearly evidenced in an early work on pea root.[150] Segments from roots of pea seedling on culture on hormonal medium containg 2,4-D and yeast extract (YE) proliferated to form a tissue which comprised 2n, 4n, and 8n cells. In such a culture, after 1 week there was an abundance of 4n cells. When YE was not included in the nutrient medium, the entire population was predominantly diploid. The effect of YE could be simulated by including cytokinin into the medium.

Also of interest in this context is the recent finding of differential responses of leaf cells of *Fagopyrum* to different growth substances stimulating cell division.[104] High 2,4-D (5 mg/ℓ) and low kinetin (0.1 mg/ℓ) stimulated division in the layer between the palisade and spongy parenchyma tissue after 72 hr. Low 2,4-D and low kinetin (0.1 mg/ℓ each) induced division primarily in spongy parenchyma cells. Whereas on a high benzylamino purine (10^{-5} M) and low IAA (10^{-6} M) medium, the divisions were localized to the palisade layer.

5. Physical Factors

Very little attention has been given to physical factors such as light, temperature, and humidity in the induction of cell proliferation. When cells are induced to proliferate, either in response to chemical environment or in response to wound reaction, to begin with, cells in the outer region alone are induced to divide. Although cell proliferation is possible in the dark, the importance of light in this process can be seen in a work on tuber tissue of Jerusalem artichoke. When explants were isolated in ordinary light, only half of the peripheral cells divided but when the explants were taken in low-intensity green light, almost all the cells of the outer region divided.[172]

Rarely, light alone can bring about callus formation. Roots from seedling of *Dioon edule* on culture on mineral-sucrose medium alone developed massive calluses in light.[160] Initiation of callus and its continued growth was due to hypertrophy of cortical cells and their periclinal divisions.[161] On transfer of these roots to dark, the tissue formed stopped growing.

C. Division Phase

Cell proliferation is followed by a rapid increase in cell number. There is almost a 1000% increase in cell number during the first 7 days of culture in explants from carrot tissue[138] and in tuber tissue of Jerusalem artichoke[173] (Figure 1). This increase in cell number is invariably accompanied by a reduction in cell size. Therefore, a dividing cell is like the cell of a meristem, a small cell without vacuoles. With a decrease in cell size the cell constituents also decrease. However, during the initial wave of division there is no decrease in RNA per cell and instead, an increase has been recorded.[137] Ultimately, a stage is reached when there is no more decrease in cell size, RNA per cell also declines, and cell constituents are in step with increase in cell number.

IV. CONTINUOUS CELL PROLIFERATION — CALLUS FORMATION

Continuous proliferation of cells is possible on regualr subculture of callus tissue, either on the medium which supported initial cell proliferation, callus induction medium, or even a simpler medium. A complex medium is rarely required.

Most of the tissues can be maintained on a mineral-sucrose medium supplemented with an auxin which serves both for induction and maintenance of the callus. On repeated subculture on maintenance medium the concentration of auxin required can be further lowered. On the other hand, there are tissues which require an auxin plus a cytokinin for induction as well as maintenance, such as pith cells of tobacco. Tobacco tissue can also be maintained on a medium containing only an auxin, but at a reduced rate. Another category is of the tissues which require a complex additive such as yeast extract, casein hydrolysate, or coconut milk for continuous proliferation. Many of such tissues are of endosperm origin.

A. Control of Callus Growth

The basic mechanism controlling unorganized growth of a callus tissue remains to be worked out. A band of microtubules encircling the nucleus in cell cortex, described as the preprophase band (PPB), is shown to predict the orientation of mitotic spindle and the plane and position of the phragmoplast and future cell plate in higher plants.[57] PPBs have not been found in unorganized tissues such as endosperm cells[35,124] and soybean callus tissue.[45] This is thought to be of functional significance

Auxin is an essential hormonal requirement for continuous proliferation of cells. It is generally a synthetic auxin such as 2,4-D. If 2,4-D is omitted, the rate of cell division is drastically reduced and cell lysis follows. True, the auxin supplied is synthetic, but cells on culture synthesize IAA[96] (Figure 2). Changes in levels of free 2,4-D and IAA in cells of sycamore *(Acer pseudoplatanus)* during a culture passage have indicated that there is rapid uptake of 2,4-D during the lag phase leading to a maximum concentration per cell on day 2, followed by a decline by day 9 (middle of log phase). The initial concentration of IAA rises slowly to a peak by day 9, then decreases rapidly by day 15 (early declining phase) and is lowest by day 23 (early stationary phase).

About the mode of action, auxins are considered to promote growth of plant cells in two ways. The first one is by cell wall acidification,[122] which causes a rapid expansion of plant cell and the second, perhaps related to the first one, is by an effect on RNA metabolism.[74] During proliferation of soybean cells induced by auxin there was rapid enhancement of RNA polymerase activity associated with nucleolus,[56] which led to the accumulation of rRNA not observed in cells starved of auxin. Also, it is shown that auxins can directly modulate the possible transcription of certain molecules.[14] This was inferred when bean and soybean cells were subcultured to fresh medium; the translatable level of a small group of mRNAs increased rapidly and then decreased to a low level after 10 hr. By starving the cells of auxin and then subculturing them into media with or without auxin, it was found that an increase in translatable mRNA for certain proteins was strongly dependent on auxin.

As for cytokinin, it has been shown that natural or synthetic cytokinin, supplied as free base to tobacco callus tissue or cell suspension, is covalently inserted into RNA sequences. Cytoplasmic rRNA species 5S, 18S, and 25S, and tRNA fractions incorporate exogenous cytokinins.[3,146,148] However, it remains to be resolved whether cytokinin insertion into RNA is related to its hormone effect.

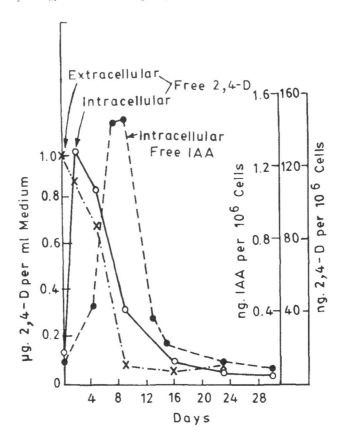

FIGURE 2.   Variations in levels of extracellular and intracellular auxins in cell suspension culture of *Acer pseudoplatanus*. (Redrawn after Moloney, M. M., Hall, J. F., Robinson, G. M., and Elliot, M. C., *Plant Physiol.*, 71, 927, 1983. With permission.)

Gibberellins are not required for growth of cells in culture, but the presence of endogenous gibberellins in cells indicates that dedifferentiated cells do produce gibberellins.[110] Their amounts are, however, much lower than in differentiated plant cells. The content of gibberellin is highest during the logarithmic phase of growth.[111] This suggests that gibberellins have a physiological role in growth of cultured cells.

The activity of enzymes, glycolytic, mitochondrial microbody, and hydrolytic, was not comparable in intact and in vitro grown tissues. Galactosidase and xylosidase were the only activities that showed a similar trend in intact and callus and cell suspension of maize.[175] Enzymes differing in activities were acid phosphatases, alcohol dehydrogenase, lactate dehydrogenase, catalase, fructofuranosidase, and glucosidase.

As for a specific enzyme group, it has been hypothesized that there is an inverse relationship between growth and peroxidase activity of plant cells.[125,153] Evidence in support of this hypothesis comes from two cell lines which differed in their growth rate.[25] A wild cell line of carrot growing less rapidly than methoteraxate resistant cell line of carrot released ten times less peroxidase than resistant cell line. It is easy to monitor the level of peroxidase, because release of peroxidase by plant cells is shown to be proportional to its synthesis.[154] Also of interest in this context is the recent finding of correlation between glyoxalase I activity and divisions of *Datura* cells.[115] This enzyme system is ascribed a regulatory role in cell multiplication including malignancy.[145] The activity of glyoxalase I increased to 184% during growth of *Datura* callus cells. In the presence of a growth stimulant, spermidine, the enzyme activity increased together with DNA and protein synthesis, whereas mitotic inhibitors, vinblastin and methylglyoxal, reduced the enzyme activity by 92 and 50%, respectively.

B. Measurement of Callus Growth

Growth of a tissue can be accounted for in terms of fresh weight increase, dry weight determination, or an increase in cell number. The last one is more reliable. An increase in fresh weight is a quick and simple method, but can be misleading, particularly when a comparison is to be made between two tisues. A tissue can be different from the other in cell size alone. Dry weight determination can also be misleading because storage products in a tissue at different times of growth can be different and can be a source of error. Therefore, a more satisfactory way of growth measurement is determination of an increase in cell number. An accurate account of number of cells in a tissue is easily possible in a friable tissue which readily disperses into free cells and small cell aggregates. By contrast, compact tissues, which are not friable can be macerated in a simpler way by a dilute solution of (5 to 10%) chromic acid or any other mineral acid at 60°C for 1 to 2 hr. Alternately, an enzyme, pectinase, can be employed for cell separation. When a comparison between two tissues is not involved the growth of a tissue can be easily measured by an increase in fresh weight.

An analysis of growth response is facilitated if measurements are possible throughout the culture period without interruption of growth. While a number of approaches have been made to meet this objective, the simplest is the growth on filter paper.[64] Plant cells are cultured on a filter paper disc resting on the surface of agar nutrient medium and growth is monitored by periodic removal and aseptic weighing of filter paper. This method allows a repeated nondestructive method of determining fresh weight increase over a period. An innovation[28] of this method is the resting of filter paper on polyurethane saturated with liquid medium. Polyurethane support allows culture of cells at a low pH. In addition, recovery of spent medium to monitor changes in pH is especially valuable for studies on mineral nutrition and metal toxicity.

C. Friable and Compact Callus

Friability or compactness of a callus is controlled by cultural conditions employed for initiation and maintenance of callus. In *Picea*[119] and *Pisum*[151] the compact callus gave rise to a friable callus, but the reverse did not occur. The friable callus was possible when concentration of yeast extract was higher relative to 2,4-D. By contrast, these two types of growth patterns were interconvertible in *Haplopappus gracilis*[16] and *Vicia faba*.[54] For the former, the level of coconut milk and naphthaleneacetic acid was crucial, whereas for the latter, it was associated with the depletion of nutrients from the culture medium.

From a compact tissue it is also possible to isolate a friable cell line by plating of cells. This was achieved in rubber tissue. The tissue of *Hevea brasiliensis*[168] on transfer to liquid medium failed to form a cell suspension; instead it broke into small pieces. However, on transfer of these smaller pieces back to agar medium after 2 months, a friable tissue was obtained which readily resulted in a cell suspension.

At times a difference is noticeable in two types of tissues. The compact callus has meristematic regions, the meristemoids. This is an indication of organization. However, meristemoids are described for many tissues which failed to differentiate.

Of late, compactness and friability of a callus has assumed added significance. On culture of cereal embryos, compact and friable tissues arise side by side, under similar cultural conditions, and it is only the compact tissue which is capable of morphogenesis whereas friable callus is nonmorphogenic. For details see Chapter 3. Therefore, it would be rewarding to understand the reasons for compactness or friability of a callus.

V. CALLUS TO CELL SUSPENSION

A suspension culture is comprised of free cells and small cell aggregates growing in a liquid medium. It can be initiated on transfer of a callus mass to a liquid medium. A friable tissue readily disperses and gives rise to a cell suspension, whereas compact tissue grows as clumps and does not yield a good suspension. Therefore, a simpler way to obtain a good suspension is to decant only the upper layer of cell suspension, leaving behind the clumps, and consider it as starting point of cell suspension culture. In other methods, either the tissue on transfer to liquid medium is shaken vigorously employing a wrist action shaker or cell separating enzymes in a nontoxic concentration are included in the culture medium until a desirable suspension is obtained.

It is possible to remove aliquots from a suspension culture employing either an automatic pipette with an orifice or a syringe with a fine cannula which will exclude cell clumps.

A. Culture Systems

A major limitation of a liquid culture is tissue aeration. Therefore, for gaseous exchange between the culture medium and the culture atmosphere, the movement of the fluid is an essential prerequisite. This movement brings about an aeration of free cells and cell aggregates and has been achieved in a number of ways, resulting in different culture systems.

1. Shake Culture

A platform shaker is a common laboratory facility. A shaker can have either a changeable platform or, as on a fixed platform, has different clamps to hold different capacity flasks. A simple mode of shaking can be reciprocal, to and fro. However, this is not an effective mode of shaking for cultures of higher plant cells, particularly when a large culture volume is involved. For instance, for a culture volume of 60 mℓ in a 250-mℓ flask, reciprocal shaker operating at the speed of 60 to 70 strokes per minute will not be sufficient for required aeration and higher speed is harmful to cells. Therefore, for the same volume, if a circular motion is possible on an orbital shaker at the speed of 60 to 70 rpm, it is better for the growth of cell suspension. The orbital movement, throw, should be the range of 2 to 5 cm. Employing this method, starting with the suspension of tobacco cells,[99] cell suspensions are maintained.

2. Spin Culture

To obtain a larger culture volume, 4 to 5 ℓ per container, spin cultures were employed.[82,127] In an arrangement inclined at 45°, one or two bottles of 10-ℓ capacity were rotated mechanically at the rate of 80 to 120 rpm. Growth of sycamore cell suspension in such a culture was of the same order as that obtained in a 250-mℓ flask on a horizontal orbital shaker.

3. Stir Culture

For a large culture volume, aeration can also be provided by magnetic stirring, supported and rotated at 200 to 300 rpm. The system can be made more effective by supplying compressed air (5 to 10 lb/in^2) from the top through a sterile inlet.[91,92,94,167]

Aeration of a suspension culture is also possible by a supply of compressed air alone.[152] A more effective way of air supply is to provide it from the base. As the air enters the vessel it expands into a large bubble which can be equal to the diameter of culture vessel if it is a long vertical column. Such a system was devised by Kurz.[81] The gradual upward movement of the bubble brought about vibration and aeration of entire culture. Vibrations additionally help to reduce cell aggregation.

4. Rotary Culture

Rotary culture system involves a slow rotation of cultures, 1 to 2 rpm, mounted on a

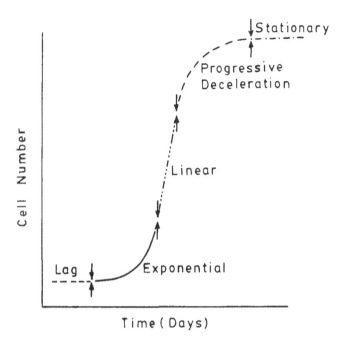

FIGURE 3. Growth curve of a cell suspension culture, grown as a batch culture. (From Wilson, S. B., King, P. J., and Street, H. E., *J. Exp. Bot.*, 21, 177, 1971. With permission.)

circular wheel-type platform.[136] The cultures/vessels are either T-tubes (tumble tubes) for a small culture volume or N-flasks (nipple flasks) for larger culture volume. Due to rotary motion, along the wheel, the cell suspension is alternately exposed to air and is immersed in a liquid, permitting a high rate of gaseous exchange.

This type of rotary culture system[136] described as auxophyton, was basically devised for study of embryogenesis in cell suspension cultures of carrot. Since then, it is widely employed for this purpose and rarely for multiplication of cells.

B. Measurement of Growth of a Suspension Culture

Growth of a cell suspension culture can be monitored by (1) cell counting or (2) determination of biomass, either by knowing total cell volume (packed cell volume) or by net increase in weight, fresh or dry. Packed cell volume (PCV) can be determined by transfer of known aliquot of cells to a graduated centrifuge tube, centrifuging the contents at 2000 *g* for 5 min. It is expressed as milliliter of cell pellet per milliliter of culture.

Cell counting is the most reliable and widely accepted method. This can be done as described for callus. An account of increase in cell number with time indicates that to begin with there is no increase. During this lag phase cells possibly prepare for division. This is followed, for some time, by an exponential phase of growth when a small fraction of cells divide. Then there is linear increase in number of cells, the log phase. This is, however, not maintained for a long period and is followed by a slowing of cell division rate possibly due to depletion of nutrients, progressive deceleration stage. Finally, a stage comes when there is no more increase in cell number, stationary phase (Figure 3).

During a culture one can also notice certain morphological features which are characteristic of each stage. Dividing cells are circular and dense, more like meristematic cells than the stationary phase cells which are elongated and vacuolar with thick cell wall.

On subculture, the stationary phase cells can be induced to divide. The duration of lag phase depends upon how long the cells have been in the stationary phase. There is practically no lag phase if the cells are taken during the linear phase of culture and subcultured to start a new culture. However, a more important factor which determines the duration of lag phase is density of cells taken at the start of culture.

As for the exponential phase, a plotting of $\log_{10}$ of growth parameter (X) against time (Y) gives a straight line. The duration of the exponential phase in an ordinary culture is hardly for four to five cell generations. This can be seen in culture of *Galium* cells, where specific rates of increase in dry weight and cell number were equal between 2 to 5 days of culture.[166] The exponential phase, however, is extendable by (1) reducing the size of inoculum,[51,103] (2) increasing the concentration of limiting nutrient,[84] and (3) subculturing cells at more frequent intervals.[9]

C. Concepts of Minimal Cell Density and Conditioning of Medium

A certain number of cells, minimal cell density, is required for growth on a defined medium. This is because the growth of cells is a function of an equilibrium between an inflow of substances from the medium to cell and an outflow of substances from the cells to the medium. In a situation where there is a small number of cells in a large volume this equilibrium is not established and consequently the cells fail to divide. However, if the cells are more than the required number, the lag phase is reduced. The reverse is also true; if the cells are less than the required number, there is a prolonged lag phase.

It is possible to grow cells at a low density if the medium is taken from a culture in which cells have been growing. This indicates that cells during their growth due to release of metabolites alter or condition the medium in such a way that it can support the growth of new cells at a low density. There is no specificity in conditioning; medium from one culture can serve to support the growth of other cells.[10]

Employing a conditioned medium it has been possible to reduce the minimal effective cell density by a factor of at least ten in sycamore cell suspension culture.[141] The conditioned medium can be prepared employing either a dialysis tube or a pyrex sinter (Figure 4A). Cells at high density are grown either in dialysis tube or pyrex sinter and this is in turn suspended in a defined medium having cells at a low density. In such an arrangement, cells can grow at a low density.

Conditioning of a medium is a complex process. It is not only possible by substances which go into solution and can stand autoclaving, but also by gaseous substances. Cells growing at high density release some volatile substances which can support the growth of cells at a low density[142] (Figure 4B). This is not possible if the gaseous substance is absorbed.

VI. CONTINUOUS CULTURE OF CELL SUSPENSION

Suspension cultures described above are limited by a finite volume of nutrient medium. Because of this, a stage comes when, due to depletion of nutrients, the cells stop dividing and enter into stationary phase. Such a culture that operates for a limited period is described as batch culture. However, batch culture can be modified to an infinite volume by either periodic or constant addition of nutrients and in this way it is possible to establish a continuous culture of plant cells.

A. Concept

To establish a continuous culture of plant cells, larger culture vessels employed are either a round bottom flask,[94,167] or an inverted Erlenmyer flask,[157] or even a big bottle.[140] Each vessel has a wide mouth and a glass lid with ports for inlet and outlet.

Aeration to culture is provided by teflon-coated magnetic stirrer supported and rotated (200 to 300 rpm) on a small glass rod at the bottom of vessel. It is supplemented by entry

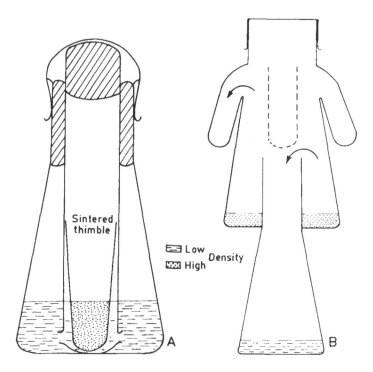

Sintered
thimble

Low
High Density

A

B

FIGURE 4. Growth of cells at low density on conditioned medium. (A) Chemical conditioning. (B) Gaseous conditioning. (A, From Stuart, R. and Street, H. E., *J. Exp. Bot.*, 20, 556, 1969. With permission.) (B, From Stuart, R. and Street, H. E., *J. Exp. Bot.*, 22, 96, 1971. With permission.)

of compressed air, through a vapor trap, a carbon filter, and two microflow miniature submicron line filters. The sterile air finally enters in the middle of culture medium via sintered glass aerator.

Initially, a continuous culture is also run as a batch culture. To monitor its growth a regular sampling is done. This is effected by a closure of air outlet leading to a buildup of pressure in the culture and this results in the flow of culture into sample receiver. The air outlet is then opened, the culture sample is drained, and then the sample receiver and line are washed with sterile water. This batch culture (Figure 5), before it enters the deceleration phase, can be modified into a continuous culture by addition of new medium.

A continuous culture can be a close continuous system if spent medium (the medium which is incapable of supporting the continued growth of cells) is siphoned out and a new medium added. In this way, a prolongation of exponential growth is possible. This drain and refill technique, however, does not permit for the steady state of growth and to achieve it, the system has to be operated as an open continuous system (Figure 6) involving a regulated input of medium and a harvest of an equal volume of culture. For the controlled supply of medium, a peristaltic pump may be employed. However, the peristaltic pump has a rubber tubing which can easily burst resulting in infection. Hence a micrometering pump is preferred. Medium input is regulated by a constant level device, the electrodes of which operate the solenoid valve and help in the harvest of culture.

The continuous culture system can be modified for an automatic sampling by inserting a needle which is linked to a multiset time clock and a fraction collector. Additional modifications are also possible such as insertion of thermometer, glass electrode, and oxygen electrode to monitor temperature, pH, and oxygen supply. It is even possible to modify a continuous culture system for monitoring the density of cell suspension.

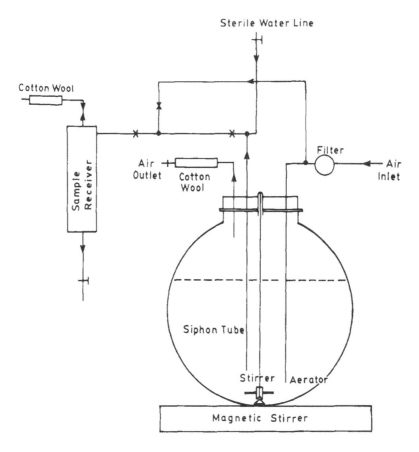

FIGURE 5.   Diagrammatic representation of 4-ℓ batch culture unit which can be adapted for continuous culture. (After Wilson, S. B., King, P. J., and Street, H. E., *J. Exp. Bot.*, 21, 177, 1971. With permission.)

B. Steady State

In a continuous culture system it is possible to achieve a steady state of growth. This concept has emerged from work on bacterial cultures.[97] In order to understand the phenomenon it will be in order to follow the microbial work. When a bacterial culture is subjected to constant dilution, by adding new medium, eventually a stage is reached in terms of continued growth and it is the manifestation of self-regulating equilibrium between organismal growth and environmental parameters.

Steady state in a bacterial culture is achieved on a principle that the cell division rate or specific rate of growth (u) is proportional to the concentration of a limiting substrate (S) in the medium. Therefore, u α S.

Further it is theorized that:

1.   As long as S is not limiting, u will be exponential.
2.   If a culture is diluted at a particular rate D, as long as u is greater than D, biomass will increase.
3.   With an increase in X, S will decrease since it is a limiting nutrient.
4.   With decline in S, u will decline.
5.   If S can be made constant, then u can also be made constant.
6.   S can be made constant by fixing the rate of dilution, D.

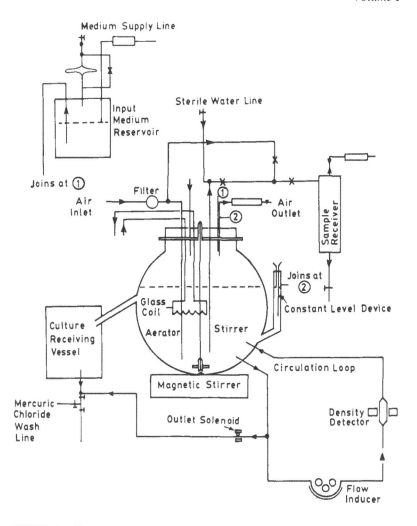

FIGURE 6. Diagrammatic representation of open continuous chemostat culture system. (After Wilson, S. B., King, P. J., and Street, H. E., *J. Exp. Bot.*, 21, 177, 1971. With permission.)

Therefore, in essence, by adjusting the rate of dilution or concentration of a limiting nutrient in the input medium one can achieve steady state provided other factors are not limiting.

The steady state in an open continuous culture system can be achieved either on the principle of a chemostat where there is fixed rate of input of new medium or a turbidostat where the input of new medium is based on optical density of cell suspension.

1. Chemostat

Steady state in a chemostat is achieved by fixing the dilution rate. In a number of cell suspension cultures, based on a variety of limiting nutrients, steady states have been achieved.

To illustrate the phenomenon, the details are given of a culture of sycamore cells in a urea-limited culture. The culture medium employed for culture of sycamore cells, after supporting an exponential growth, normally becomes limiting in urea. If the concentration of urea in an input medium is increased by a factor of three one can see the adjustment by the culture in terms of cell number by this single change in nutrient concentration. Both cell number and dry weight become constant (Figure 7). An indirect support to the concept of

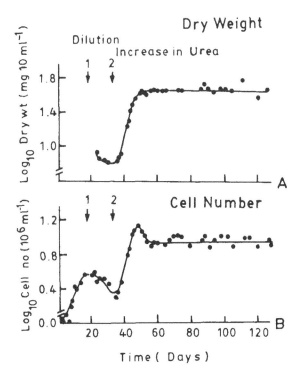

FIGURE 7. Steady state in a chemostat culture of *Acer* cells maintained on urea-limiting nutrient as sole nitrogen source. Dilution began at arrow 1 with medium containing 6.6×10^{-3} *M* urea. At arrow 2 the urea concentration was increased to 19.8×10^{-3} *M*. (From King, J., *Plant Sci. Lett.*, 6, 409, 1976. With permission.)

steady state is through perturbation experiments. In a nitrate-limited chemostat culture of sycamore cells, a stepdown in the concentration of NO_3 resulted in oscillations in growth, intracellular nitrogen pool, and extracellular NO_3 concentration.

2. Turbidostat

In a turbidostat, steady state is achieved by fixing the value of biomass. When biomass increases beyond a fixed limit, a control device operates and the new medium is added to the culture to dilute the biomass to the required value. The main utility of a turbidostat is that it can serve as a constant source of exponentially growing cells which can be utilized for fundamental or applied research.

C. Continuous Culture at Industrial Scale

A main limitation of the different designs of culture vessels for a continuous culture of plant cells is that they are not suitable for large-scale growth to realize the goals of plant biotechnology. Also plant cells have inherent problems. They are large, 20 to 150 μm, and in culture they can assume an elongate form due to the presence of cell wall. These cells are capable of withstanding tensile stress but they are sensitive to relatively slight shear stress. Moreover, it is difficult to obtain a homogenous system for plant cells because daughter cells formed tend to stay together following cell division and, particularly towards the end of growth phase cells, tend to excrete polysaccharides, whose stickiness enhances the formation of cell clumps.

In particular, the low shear tolerance of plant cells[87] requires a new approach in design

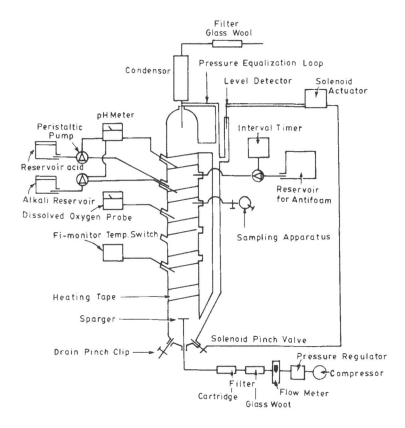

FIGURE 8. Diagrammatic representation of air-lift bioreactor system for continuous culture of plant cells. (From Smart, N. J. and Fowler, M. W., *J. Exp. Bot.*, 35, 531, 1984. With permission.)

if cells are to be cultivated on a large scale. An impeller speed of 200 rpm is sufficient to cause cell death.[32] A bubble column reactor[131] is reckoned to be an alternative. The design of an airlift column bioreactor, comprised of a 100-mm-bore-diameter glass tube of 1.27 m long, is shown in Figure 8. Such a system has characteristics of low shear and good mixing. This system has been operated successfully at laboratory bench scale, both as batch and continuous cultures, and subsequently successfully scaled up to 85-ℓ capacity. The growth profile of *Catharanthes roseus* cells grown in an airlift column bioreactor of 10-ℓ capacity operated on chemostat principle is shown in Figure 9. More recently,[20] cells of *Berberis wilsonae*, which produce berbarine-type alkaloids, were cultured in a 20-ℓ airlift bioreactor. It was found that formation of phenolic alkaloids (columbamine and jatrorrhizine) and berbarine were dependent on concentration of dissolved oxygen. With 50% dissolved oxygen tension in the medium, it was possible to have more than 3 g/ℓ of alkaloid content.

It is anticipated that one day the biosynthetic capacity of plant cells will be realized employing cell cultures analogous to microbial fermentations. However, in such a proposition one has to account for the characteristics of growth and metabolism of plant cells which are quite dissimilar to that of microbial cells. The basic differences are

1. Slow growth of cells.
2. Low shear tolerance.
3. Genetic instability.
4. Diploid nature.
5. Cell aggregation.

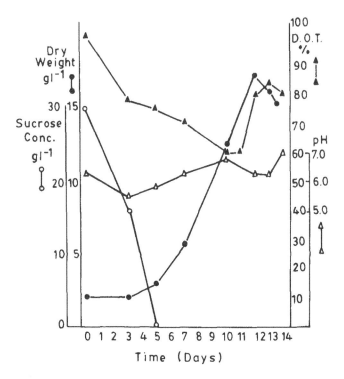

FIGURE 9.   Growth profile of *Catharanthus roseus* cells grown in a 10-ℓ capacity air-lift bioreactor system. (From Smart, N. J. and Fowler, M. W., *J. Exp. Bot.*, 35, 531, 1984. With permission.)

6.   Especial requirements for biosynthesis.
7.   Low yield of products.
8.   Products are intracellular.

Slow growth of plant cells is the basic limitation. Doubling time of different cells has been worked out to be between 20 to 60 hr. Within this period, a microbial cell produces 10^{12} to 10^{36} progeny. Slow growth introduces cost considerations for generation of biomass. Coupled with this are other factors such as low yield of product and its intracellular storage which necessitates an extraction procedure resulting in wastage of biomass. Also on continuous culture, cell lines are unstable and there is a decline in productivity. Therefore, due to the diploid and polyploid nature of plant cell, it is also not easy to raise high-yielding mutant cell lines. About aggregation and a low shear tolerance of plant cells, a mention has been made earlier. And last, but not least, certain secondary metabolites are not formed in cells on culture unless the cells undergo organization producing roots or shoots. An organized tissue is further more difficult to handle than cell aggregates. Also, the yield is not the same in a small batch culture and a large fermentor culture. This is evidenced in a recent work on nicotine production[121] by tissue cultures of tobacco. Mixotrophic green cell suspension produced up to 5.3% or 920 mg/ℓ of nicotine whereas 20-ℓ fermentor cultures accumulated 1% nicotine with a final yield of 28 mg/ℓ thus producing much less.

At the present level of technology, only high-value compounds may be considered for production in plant cell cultures. According to current estimates $1000/kg is the lowest price. At present, the only commercially viable project is the production of shikonin by cell cultures of *Lithospermum erythrorhizon*. Shikonin, a mixture of red pigments, is used in Japan as medicine and dyestuff. Its traditional source is roots and current price is $4000/

kg. From intact plants, when they are 2 to 3 years old, about 1 to 2% of dry weight shikonin production is possible whereas cultured cells start producing shikonin in 3 weeks and can be up to 14% of dry weight. This high productivity in cell cultures could be achieved in a two-stage process. First, conditions were optimized for growth of cells and then they were optimized for product formation. Thus, decoupling cell growth with product formation is one of the possibilities for improved productivity and industrial exploitation of cell cultures. This strategy was adopted for berbarine production from suspension cultures of *Thalictrum minus*.[101] Actively growing cells in the presence of 2,4-D produced little berbarine. When cytokinin was included, cell growth was inferior, but berbarine-producing activity was remarkably enhanced. The yield was as high as 20 mg of berbarine per 30 mℓ of medium in 2 weeks.

VII. CELL POPULATION

Only a fraction of cells divide at one time in a cell suspension or callus culture. This is due to the fact that in different cells, DNA synthesis and other events of cell division do not occur simultaneously. Such a population of cells in which cell division events are out of phase is described as asynchronous. By contrast, when events of cell division in different cells are in a phase, such a cell population is described as synchronous. In brief, in a synchronous cell popluation most of the cells divide at one time. A population of cells which is asychronous can be made synchronous.

A. Induction of Synchrony

Although plant cell populations, particularly those raised in vitro, are, in general, asynchronous, in a few systems synchronous divisions are reported to occur for a short period only. An example is the synchronous cell division in explants from tuber tissue of *Helianthus*.[174] Mitosis, an increase in number of cells at a given time, is taken as an index of synchrony. Other events of cell division such as protein synthesis can also serve as indices. Synchrony has been induced in asynchronous cell populations by a number of ways, described below.

1. Inhibition

Synchrony can be achieved employing inhibitors which can reversibly block any specific event of cell division. In the presence of an inhibitor, cell division is allowed to proceed up to a particular stage only and all cells are held up at this point. When the block is released, cells proceed to the next events synchronously. The evidence of synchrony has been the peaks of mitotic divisions.

The most commonly employed specific inhibitors are inhibitors of DNA synthesis. For the induction of synchrony in *Haplopappus gracilis*,[39] 5-aminouracil, hydroxyurea, and thymidine were employed. Fluorodeoxyuridine (FUDR) was also employed for *H. gracilis*[39] and *Datura innoxia*.[19] Synchronization of cells has also been achieved by blocking the cell cycle with hydroxyurea and releasing the block in the presence of colchicine as described for *H. gracilis*,[39] *Triticum monococcum* and *Petroselinum hortense*,[144] and *Datura innoxia*.[93] However, in certain cases, such as carrot and poppy, satisfactory mitotic indices could be achieved by colchicine treatment only.[88] The antibiotic aphidicolin has also been employed alone or in combination with colchicine to induce high frequency of mitotic indices.[100,59] This antibiotic is inhibitor of DNA polymerase a, and its application resulted in arrest of tobacco cells at G_1/S phase of cell cycle. Immediately after removal of chemical, DNA synthesis started and 52% mitotic index was recorded. The synchrony was due to increase in cytokinin content during G_2/M boundary.[106]

Nonspecific inhibitors employed for induction of synchrony were periodic flushing of chemostat cultures of soybean with nitrogen[29] and ethylene.[30]

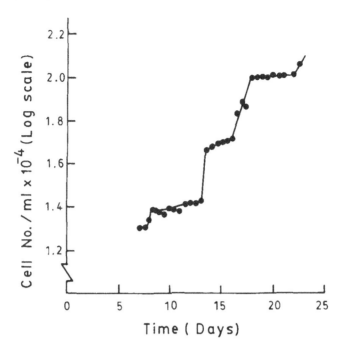

FIGURE 10.   Cell synchrony in 4-ℓ batch culture of *Acer pseudoplatanus.* (Redrawn from King, P. J., Cox, B. J., Fowler, M. W., and Street, H. E., *Planta,* 117, 109, 1974. With permission.)

2. Starvation

Synchrony has also been achieved by starvation. In the process of synchronization, the starvation of cells in terms of an individual compound has proved to be more effective. A cell culture, after dividing, exponentially enters into the stationary phase due to depletion of nutrients, and when subcultured at low density, starts dividing synchronously. Alternately, cells are subcultured for a few to several days at a low density in a medium lacking an individual compound only, and on subculture to a new medium enriched with the missing substrate there are waves of synchronous divisions.

Culture of tobacco tissue, having an obligate requirement of cytokinin, could be synchronized by withholding cytokinin for some time and the addition of cytokinin resulted in synchronous divisions.[69,68] However, auxin starvation did not result in synchrony. By contrast, cultures of carrot were synchronized by withholding 2,4-D and reculture on 2,4-D-enriched medium brought about a wave of divisions.[105] On critical analysis, starvation of 2,4-D was shown not to result in synchrony and instead produced periodic oscillations in cell number.[43] In carrot cultures synchrony was also possible by a combination of cold treatment and phosphate starvation.[109] The indices of synchrony were DNA and protein synthesis, which were periodic.

Starvation of *Acer pseudoplatanus* cells primarily due to nitrate depletion, on holding at stationary phase for about 2 weeks, resulted in synchronous divisions on subculture at low density to a medium containing nitrate.[75,52] The evidences of synchrony recorded were a stepwise increase in cell number (Figure 10), mitotic index, and DNA accumulation.

Cultures of *Vinca rosea* were synchronized by phosphate starvation. A preculture on a phosphate-free medium for 4 days and then a return to normal medium induced synchrony.[76] Similarly, cell cultures of *Catharanthus roseus* were synchronized by double starvation of phosphate.[2] After 6 days of culture, cells were transferred to phosphate-free medium at a low density (about 3×10^5 cells per milliliter) and maintained for 3 days. This was followed

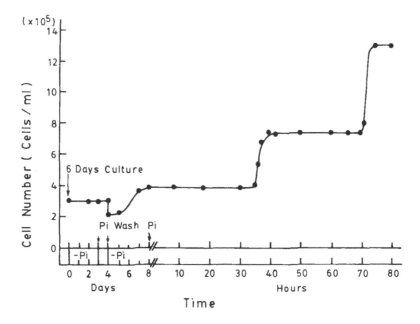

FIGURE 11. Induction of cell synchrony, in cell suspension culture of *Vinca rosea*, employing double phosphate starvation method. Pi, addition of phosphate; -Pi, culture in phosphate-free medium; wash, washing of cells with phosphate-free medium. (From Amino, S., Fujimura, T., and Komamine, A., *Physiol. Plant.*, 59, 393, 1983. With permission.)

by a pulse of KH_2PO_4 (0.2mM). After incubation for 24 hr, the cells were centrifuged and washed twice with phosphate-free medium and then transferred to phosphate-free medium at a density of 2×10^5 cells per milliliter. After maintaining for 4 days on phosphate-free medium, KH_2PO_4 was added at a final concentration of 0.675 mM (half of that available in MS medium). The second phosphate feeding induced synchronous cell division (Figure 11). Double starvation was found to be necessary to obtain well-synchronized cell population. In a single phosphate starvation only partial synchronization was obtained.

The synchronized cells are a better system for studies on cell cycle. A perusal of data in different cell systems indicates that after five to six synchronized doublings of cell number, the culture becomes asynchronous. The persistence of synchrony is unprecedented and should not be expected.

3. Starvation vs. Inhibition

Starvation of cells is considered to be less damaging than use of inhibitors for cell synchronization. In synchronization of *Haplopappus* cells on employing hydroxyurea (an inhibitor of DNA synthesis) chromatid fragments were frequently observed.[39] As for specific starvation, phosphate starvation is considered better because it prolonged the viability of *Acer* cells compared to sucrose starvation or starvation of other nutrients.[53]

VIII. CELL CYTOLOGY

A callus or cell suspension culture comprises free cells and cell-aggregates. The latter can be small aggregates of 30 to 100 cells in a friable tissue or big clumps in a compact tissue. Among free cells, most are spherical with a small proportion of tubular cells. Occasionally, giant cells, spherical, or abnormally elongated tubular cells are seen. At times, cells are of diverse shapes.

Cell shape is basically a function of culture duration and nutrient conditions. When a tissue is not regularly subcultured, formation of abnormal cells is favored. It is more pronounced if the cells are retained in the stationary phase for a long period. A frequent subculture of suspension culture, at a 2 to 3 day interval, helps maintain actively growing small circular cells of 30 to 100 μm. Nitrogen is the key element affecting the shape of sycamore cells. When cells were maintained on a medium containing nitrate as a sole source of nitrogen, spherical cells were formed, and when organic nitrogen (casein hydrolysate, urea, and cystein) was given excessively elongated cells were formed.[128]

A. Changes in Number of Chromosomes

Cells in culture, either as callus or suspension, are unstable in chromosome number. The variability may appear as early as the first subculture;[58] otherwise it is normally the feature of a long-term culture.

The types of variation recorded are polyploidy, an increase in chromosome number which is a simple multiple of a basic number, and aneuploidy, an increase in number which is not a simple multiple of a basic number. Endomitosis, chromosome duplication within intact nuclear membrane, and endoreduplication, duplication of chromatids of each chromosome during metaphase, are the principle causes of polyploidy.[33,34] In endomitosis there is formation of a restitution nucleus, due to failure of spindle formation.[8] Another reason for polyploidy can be fusion of spindles during synchronous divisions.[95] Polyploid cells of odd series (3n, 5n, etc.) may arise through nuclear fusion[143] or genome segregation during polyploid mitosis[34] or even nuclear fragmentation associated with first division of callus initiation.[27] Aneuploidy may result from translocation, amitotic nondisjunction, or deletion.[41]

B. Changes in Karyology

Not only a change in number of chromosomes but karyotypic changes have also been noticed in cells on culture. This has been documented in *Crepis capillaris*[123] and *Haplopappus gracilis*,[130] due to the small number and large size of chromosomes. In *C. capillaris* three pairs of chromosome can be identified. The first comprises long chromosomes (L), in another pair each chromosome has a satellite (SAT), and the third pair is of small chromosomes (S). Several abnormal karyotypes were seen in callus cultures. An analysis revealed that distribution and rearrangement was not random and one particular region of SAT chromosome was involved in chromosome breaks.

In *H. gracilis* deletions were more frequent in the long arm of chromosome II. In aneuploid cells and extra chromosome appeared more often than chromosome II. This indicates that genetic material on chromosome I has greater survival value under in vitro conditions. In addition, microchromosomes and ring chromosomes have been reported in this genus.

Employing a more resolving technique like giemsa C-banding it was possible to demonstrate how dramatic the internal rearrangements can be in *Crepis capillaris*.[4] The rearrangement in culture was so extensive that it was difficult to relate the chromosomes from cells in culture to root tip chromosomes.

In addition to these genera where karyotypic analyses have been possible abnormal chromosomes with structural alterations have been recorded in many genera. Outstanding examples are dicentric chromosomes in wheat, *Triticum monococcum* and *T. aestivum*,[72] and *Daucus carota*.[9] A unique feature was the presence of megachromosomes in wheat.[72]

C. Factors Affecting Cytological Changes

Cytological stability for a short period can be achieved by selection and use of appropriate culture medium[102] and by having short intervals between subcultures.[40] However, widespread occurrence of cytological instability of cells in a culture necessitates an understanding of factors affecting this response. These are given below.

1. Polysomaty

It is an initial cause of polyploidization.[33] Polysomaty is the heterogenous population of cells, variable in chromosome number, in an explant employed for raising a tissue. Polyploid cells in vivo do not divide possibly for want of special nutrient conditions but divide in vitro because such conditions are met.

2. Nutrients and Hormones

The composition of nutrient medium is another important factor. It is shown in an early work on induction of proliferation of pea root segments.[150-151] When auxin 2,4-D alone was employed, as a chemical stimulus for cell proliferation, diploid cells divided, but when 2,4-D was employed along with yeast extract or kinetin tetraploid cells were selectively favored for proliferation. Kinetin (K) is possibly a major factor in inducing polyploidy in an established callus. Haploid cell suspension of *Atropa belladonna*[116] showed increased polyploid mitoses even in second passage on medium containing K 10^{-5} *M* and NAA 10^{-5} *M* (Figure 12), whereas it was predominantly diploid even up to fifth passage on medium containing K 5×10^{-7} *M* and NAA 10^{-5} *M*. Similarly, K and YE increased the frequency of polyploid mitoses in *H. gracilis*.[129] However, in contrast to these results, a high cytokinin:auxin ratio helped maintain haploidy of *Datura*[155] cells.

In an established callus or cell suspension, type and concentration of auxin, which is an essential factor for growth of cells, are important factors controlling cell cytology. Of the auxins, 2,4-D is considered to favor polyploidization either by induction or favoring the growth of polyploid cells. Compared to 2,4-D, NAA is relatively less harmful. The tissue of *H. gracilis* changed from diploid to wholly tetraploid within a period of 6 months and the change was much slower if NAA was employed instead of 2,4-D.[143] Also in *Vigna sinensis* 2,4-D caused more chromosomal variation than NAA.[65]

In some tissues, it has been found that 2,4-D or NAA at lower concentration selectively favored polyploidization which could be controlled by employing higher concentration. Tissue of *Helianthus annuus* showed an increase in polyploid mitoses on medium containing 0.25 mg/ℓ of 2,4-D and at 20 mg/ℓ it favored diploid mitoses.[70] Similarly, seedling callus of *H. gracilis* remained predominantly diploid up to 80 days on medium containing auxin NAA (1 mg/ℓ) and cytokinin. However on this medium it progressively became polyploid which could be arrested by increasing the level of NAA to 4 mg/ℓ.[11]

3. Genotype

Finally it is worth recording that there are tissues which show cytological stability. These are tissues of *Helianthus tuberosus*,[112] *Dendropthoe falcata*,[66] and *Lilium longiflorum*.[126] Such a tissue, in a way, is a better material for cytological studies because the intrinsic role of material, if any, is ruled out.

A close relationship of genotype to in vitro chromosomal variability is seen in *Apium graveolens*[22] and *Nicotiana tabacum* su/su genotype.[42]

4. Original Ploidy of Cells

A comparative study of haploid and diploid tissues, of *Crepis capillaris* in long-term cultures indicated that diploid cells are more stable than haploid cells. Diplodization of haploid cells was more common than occurrence of tetraploids in a diploid cell line.[123]

IX. CELL VIABILITY

An account of growth of a cell suspension reveals that not all cells divide, indicating that not all cells are alive. This is particularly evident when a culture is maintained for a long period in the stationary phase before subculture to a new medium. The viability of plant

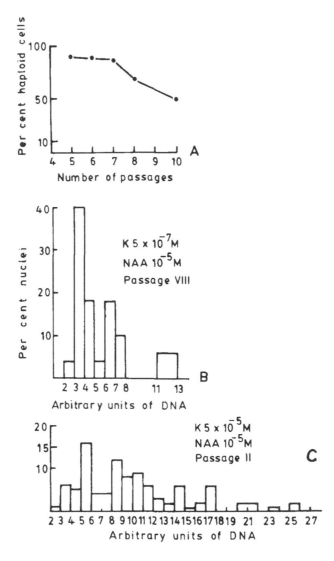

FIGURE 12.   Ploidy level of haploid cell suspension culture of *Atropa belladonna*. (A) Percent haploid cells on successive subcultures. (B) Microdensitometer profile of cells on standard medium after VIII passage. (C) The same, after II passage cells maintained on higher level of cytokinin. (From Rashid, A. and Street, H. E., *Plant Sci. Lett.*, 2, 89, 1974. With permission.)

cells can be tested by a number of methods. However, not all tests are considered to be reliable. The tests which are particularly misleading are staining reaction of living cells to tetrazolium chloride and staining reaction of dead cells to phenosafranin. Given below are tests which have given consistent results.

A. Staining with Vital Stains

Fluorescein diacetate (FDA), fluorescein isothiocyanate (FITC), rhodamine diacetate (RDA), and rhodamine isothiocyanate (RITC) are vital stains which are frequently used to test the viability of cells.

In principle, these compounds are nonfluorescing and nonpolar and are readily permeable across the plasma membrane. However, inside the cell they are broken by esterases into a fluorescent polar portion which is unable to permeate through plasma membrane. In viable cells with an intact plasma membrane the accumulation of polar compound can be detected under UV light in the form of fluorescence, which is bluish green in FDA and FITC and red in RDA and RITC.

For staining, a saturated solution of one of these vital stains is prepared in acetone and a drop of this solution is added to 1 to 2 mℓ of cell suspension. After about 10 to 15 min at room temperature the cells are examined employing a fluorescence microscope. In bright field illumination all cells are visible and in dark field illumination when exposed to UV light only viable cells showing fluorescence are seen. Dead and broken cells, since they cannot retain fluorescein, do not show any fluorescence and are invisible.

B. Exclusion of Evan's Blue

Of the reliable tests for cell viability a relatively simple method is staining with Evan's blue. This stain (0.025% solution) is taken up only by dead and broken cells. Live cells exclude it. This test is simple and can be performed employing a light microscope.

Several workers have reported chemical or heat killing experiments to establish that specific stains distinguish living from nonliving cells. However, a quantitative relationship between these stains and senescing tissue is not available. Also quantitative information is lacking between staining properties of cells and their potential to divide. Staining properties of rose cells during senescence did not correlate with their ability to grow because an increasing proportion of living cells appeared to lose their ability to divide as senescence progressed.[133]

X. CELL SENESCENCE

Regulation of senescence in an entire plant is associated with changes in the balance of senescence-promoting hormones, ethylene and abscisic acid, and senescence retarding hormones, auxins and cytokinins. Essentially, deficiencies of auxin and cytokinin also trigger senescence of cells in culture.

Deprivation of auxin alone resulted into senescence of *Pyrus communis* cells, which are cytokinin autonomous. However, senescing cells could be kept quiescent, protracted senescence, by including mannitol in the auxin-deficient medium. In this way cell death began after 14 to 16 days. The ability of cells to divide again in presence of 2,4-D was retained up to 9 days but was rapidly lost thereafter.[31] In senescing cells a sharp increase in protein and RNA syntheses was observed which appeared to be directly involved in cell death. It is proposed that auxin deprivation unmasks the synthesis of new proteins involved in cell death. This transient increase in protein synthesis, preceding cell death, was affected by cycloheximide and it also changed the timing of cell death. Cycloheximide also induced an alternate respiration pathway which was cyanide resistant.[120] This corresponds with CRR pathway of senescent fruit cells. Another change beginning around day 9 was an efflux of three amino acids, serine, threonine, and aspartic acid, of which serine represented more than 52%. However, exogenous supply of serine to medium did not have any senescence-promoting effect.[7]

An actively growing cell suspension of pear produces ethylene, which decrease rapidly when cells enter into the stationary phase. When cells are deprived of auxin, division gradually ceases and ethylene production falls to a barely discernible level. However protracted senescent cells of pear, on deprivation of auxin and on culture on mannitol supplemented medium, produced a large quantity of ethylene. This was more apparent on subculture to auxin medium.[114] This was also possible in response to other stimuli such as addition of $CuCl_2$ or 1-aminocyclopropane-1-carboxylic acid (ACC). Maximum rates of production of ethylene were achieved about 12 hr after the addition of IAA, NAA, or 2,4-D. From these results it can be concluded that auxin depletion initiates senescence, which in turn leads to a transient increase in inducible ethylene production and eventual cell death.[113] Prior to senescence cells exhibit a transient increase in protein synthesis, an increase in leakage of serine, and an increase in alternate electron transport pathway. These transitions are close counterparts to events in senescent plants and tissues and imply that the cells in culture constitute a comparable system.

Also cell suspension of *Lupinus polyphyllum*[169] at the start of stationary phase formed about 100 mmol/ℓ of ethanol of which 50 to 70% was released into the culture medium. The activities of extracellular peroxidase, oxidase, acid phosphatase, esterase, and protease increased during active growth and decreased during stationary phase when ethanol formation began.

Prevention of senescence of rose cells, at a stationary phase of growth, was possible by diamines (1mM cadaverine or putrescine) or polyamines (0.1 μM sperimine or spermidine) along with 2% sucrose.[98] Senescence of cultures was assayed by microscopic examination of cell aliquotes removed at 10-day intervals and treated with FDA.

XI. FREE-CELL CULTURE

Culture of free cells is an indispensible technique for isolation of cell clones. Culture of isolated cells is not only essential for an insight into their physiology but also for realization of natural or induced variations.

To obtain a free-cell culture it is essential that cell-aggregates are removed, because it has been observed that in a cell suspension the growth is mainly due to division of cells which are the units of an aggregate and it is rare that free cells divide. Free cells of wheat failed to divide in a suspension culture at any cell density.[159]

A. Cell Plating

A simple technique for culture of free cells and isolation of individual cell clones is the plating of plant cells on agar,[12] analogous to a microbial culture.

In this technique, a suspension of free cells is mixed with molten agar at 40 to 42°C. The contents are dispensed in a petri dish in such a way that cells are evenly distributed and embedded in a thin layer of agar when it forms a gel at room temperature. The position of an individual cell can be marked by examining the inverted dish under a microscope. To prevent infection the dish is sealed with parafilm or cellotape. In order to have a known number of cells per milliliter of plating medium, the cell number is determined in the cell suspension. This suspension is adjusted, either by dilution or concentration, to have twice the number of cells than required per milliliter and then mixed with agar medium in a ratio of 1:1. The agar medium should also have twice the concentration of agar and nutrients required.

Plating of plant cells permits a screening of large number of individual cells for isolation of cell clones. This technique has been successfully employed for raising of cell clones in many plants.

Innovation of this technique involves placing a layer of nutrient agar at the bottom of petri plate, often referred to as bottom agar or bottom layer, and layering of cells over it in a thin agar, often called soft agar or cell layer. This innovation is particularly helpful when cells take a long time to divide. If cells are plated on a thin layer of nutrient agar and take a long time to divide these are likely to die in this interval due to drying up of the medium. However, in this technique of having two layers of agar it is often difficult to locate individual cells, unless a highly purified grade of agar is employed.

B. Plating Efficiency

By knowing the total number of cells per plate and the number of colonies formed by them it is possible to calculate plating efficiency (PE).

$$PE = \frac{\text{Number of colonies per plate}}{\text{Number of cells per plate}} \times 100$$

Number of colonies per plate are counted on a colony counter. Alternatively, a shadow graph print is made of the plate by resting the plate on a photographic paper. While exposing the plate, a negative of a ruled area is placed in the negative holder of enlarger. In this way, this grid is superimposed on the photograph. In the photograph the colonies are counted employing $10\times$ magnification. The average number of colonies in 5 big squares or 20 small squares is good enough to give an overall number of colonies per plate.

1. Need for Increased Plating Efficiency

The principal aim of cell plating is the isolation of cell clones. The colonies formed should be wide apart which does not allow a mixup of cells and facilitates an easy separation of colonies formed. However, this ideal situation is difficult to achieve because in order to have good plating efficiency (PE) cell densities of the order of 10^4 to 10^5 cells/mℓ are required. At these high densities the distance between two cells is hardly more than three to four times the diameter of a cell, and it is almost impossible to separate colonies. Hence, a low cell density and high PE should be the aim of experiments on cell plating.

C. Factors Affecting Plating Efficiency

The growth of a cell into a new medium is the function of an equilibrium between the cell and its new environment. On transfer to a new medium the cells, due to their permeable nature, lose essential metabolites to new medium. At a low cell density this loss may far exceed the biosynthetic capacity of cells for replacement. Therefore, growth is possible only when cells acquire higher biosynthetic potential or become more effective in utilizing the nutrients from the medium. Accordingly, a number of factors affect PE.

These factors are density of cells, composition of medium, levels of growth hormones, and other growth adjuvants like coconut milk, yeast extract, casein hydrolysate, and physical conditions like light, temperature, and gaseous atmosphere of culture vessel (Figure 13). In addition, the age of inoculum at the time of plating also affects PE.[38,89] Therefore it is advisable (1) not to take cells which have been in stationary phase for a long period and (2) not to expose the cells to a temperature beyond 35 to 40°C during their mixing with molten agar.

1. Light and Temperature

In a consideration of light, affecting PE, one has to account for the types of cells, chlorophyllous or achlorophyllous.

Light is certainly inhibitory to colony formation by cells which are achlorophyllous. Instead, an incubation in dark is favorable. Colony formation by sycamore cells was relatively inhibited in those plates which were periodically exposed to low-intensity light for microscopic examination than those plates which were kept in the dark.[139]

By contrast light is promotory to colony formation by chlorophyllous cells of *Nicotiana*.[13] There is an increase in PE with an increase in intensity of light. Since the effect was possible in blue as well as red light it is attributable to photosynthesis providing additional energy. It appears that the response to light is a genetic effect, even for chloroplast-free cells of *Nicotiana* light is promotory for colony formation.

As for temperature, most workers routinely incubate cells at 25°C. However, in *Haplopappus*[38] at 25°C only aggregates of five to ten cells divided and for growth of free cells optimal temperature was 30°C. A further increase to 35°C was harmful.

2. Nutrients

The most important factor affecting the growth of free cells is the nutrient medium. Most

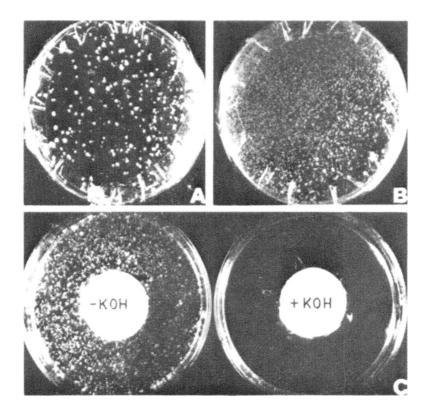

FIGURE 13.   Plating of free cells of *Indigofera* on medium with (B) or without casein hydrolysate (A) and with or without CO_2-atmosphere (C). (From Bharal, S. and Rashid, A., *Biol. Plant.*, 26, 202, 1984. With permission.)

workers[12,15,17,37,50] have employed media fortified with undefined substances like coconut milk, yeast extract, or casein hydrolysate. However, the goal should be the culture of free cells on a defined medium. Of interest, in this context, is the report of the first successful culture of free cells of *Haplopappus* and *Vitis* on defined medium.[118] The success in *Haplopappus* was, however, limited to 2 out of 58 cultures, whereas cells from crown gall tissue of *Vitis* grew at an efficiency of 40 to 70%. This may be due, in part, to the greater biosynthetic capacity of tumorous cells which are hormone autonomous. For cells of *Haplopappus*,[38] the casein hydrolysate and coconut milk requirement could be replaced by increasing the levels of pyridoxine, thiamine, and sucrose and also by increasing the concentration of phosphate and deleting potassium iodide from the medium. Also for cells of *Convolvulus*[36] yeast extract could be replaced by kinetin and glutamic acid.

The role of amino acids in growth of free cells has been elaborated in a work on green and nongreen cells of *Nicotiana*.[85] In order to obtain high PE at the low cell density of green cells, the essential requirement of coconut milk could be met by inclusion of kinetin and glutamine. The requirement for glutamine is specific; it could not be replaced by glutamic acid or asparagine. The specific requirement of glutamine and kinetin were there only when cells were grown at densities lower than 500 cells per milliliter indicating thereby that at higher densities these cells are able to synthesize glutamine and cytokinin in sufficient quantity for growth.

As for nongreen, heterotrophic cells of *Nicotiana*, glutamine and kinetin could not support their growth at a low density (320 cells per milliliter). In order to grow these cells at this density the medium had to be modified by inclusion of conditioned medium and glutamine

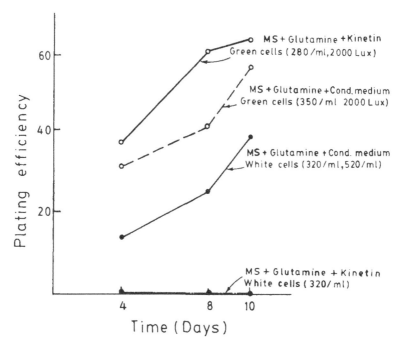

FIGURE 14. Effect of conditioned and defined media on PE of chloroplast-free and chlo-rophyllous cells of tobacco. (From Bergmann, L., in *Plant Tissue Culture and its Biotechnological Applications,* Springer-Verlag, 1977, 213. With permission.)

(Figure 14). An analysis of conditioned medium revealed that it had sufficient cytokinin but an inadequate glutamine level. Further it is interesting to know that kinetin and glutamine were sufficient to support the growth of green cells which were not washed prior to culture. This points out (1) differences in the synthetic capacity of green vs. nongreen cells, (2) carryover effects of the earlier medium to which cells were grown, and (3) presence of metabolic pools of amino acids in cells. In a follow up of changes in amino acid content of cells it was found that there was a sharp fall if cells were transferred to nitrate medium than medium containing ammonium and nitrate[13] (Figure 15).

Notwithstanding the success described above about the culture of free cells at low density on complex medium or conditioned medium the aim should be growth of free cells on a defined medium. Noteworthy in this context is the devise of a defined medium which could support the growth of 25 to 50 cells per milliliter of *Vicia hajastana.*[71] In addition to minerals the medium comprised glucose, sucrose (as main carbon sources), a mixture of 8 other sugars and sugar alcohols, a mixture of 14 vitamins, 2,4-D, zeatin, NAA, glutamine, alanine, glutamic acid and cysteine, a mixture of 6 nucleic acid bases, and a mixture of 4 organic acids of TCA cycle. The nutritive value of undefined substances in terms of growth factors for free cells can be understood by the finding that on inclusion of casamino acids and coconut milk to this medium it was possible to grow cells at a density of 1 to 2/mℓ.

3. Gaseous Atmosphere

Of late, CO_2 has been recognized as an essential factor for the growth of free cells, irrespective of whether they are chlorophyllous or achlorophyllous. There is significant promotion in colony formation in an atmosphere enriched with CO_2. How CO_2 affects the growth of nonchlorophyllous cells is not clear. Nonchlorophyllous cells of sycamore failed to form colonies at a low density when they were grown in CO_2-free atmosphere, whereas colonies appeared when air containing 1% CO_2 was given.[47] Similarly, cells of *Indigofera* failed to form colonies in a CO_2-free atmosphere (Figure 13C) at a fairly high density of

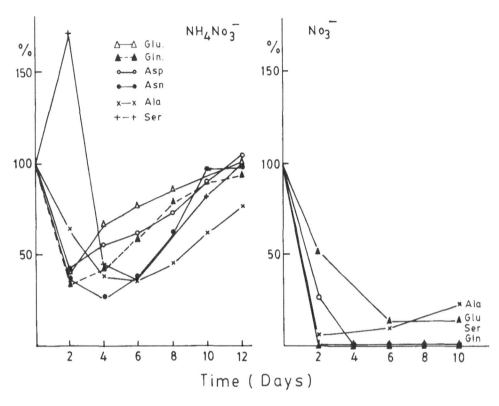

FIGURE 15. Changes in free amino acid composition of tobacco cells grown as suspension culture with different nitrogen sources. (From Bergmann, L., in *Plant Tissue Culture and its Biotechnological Applications,* Springer-Verlag, 1977, 213. With permission.)

2.5×10^3 cells per milliliter. Even at a still higher density of 5×10^3 cells per milliliter a few colonies appeared, in a CO_2-free atmosphere, which were slow growing.[15]

4. Nurse Culture or Feeder Layer Support

A less common mode of culture of free cells is through nurse culture. In a paper raft technique free cells are dispensed over a filter paper which is in turn placed over a callus tissue. In this way nurse callus is able to support the growth of free cells, at a low density.

In a different way, nurse tissue was employed to culture free cells of pea root callus.[149] Free cells were placed around the outer part of agar medium and in the center was placed the nurse callus. About 8% of the free cells divided to form colonies.

Of late, attempts have been made to support the growth of cells at low densities by cross-feeding them with a feeder layer of actively growing cells at high density or even nondividing cells at high density. To begin with, this technique was employed to isolate mutants from *Datura* cells plated, at low density[162] and supported by feeder layer of cells of *Glycine max.* The method was more revealing when about 75% of the free cells of *Haplopappus*[63] could grow on filter placed onto the feeder layer of irradiated *Haplopappus* cells. Also, cells of *Zea mays*, when plated at low density, grew into callus colonies within 2 weeks with the assistance of a feeder layer of maize suspension cells. Further it was found that cell density of feeder layer was critical (Table 1) to the satisfactory feeding of plated maize cells.[132] If the concentration of the cells in the feeder layer was below the minimum initial cell density for the suspension culture, the plated cells did not grow. Little information is available about the identity of beneficial substances provided by nurse/feeder cells. Protoplasts in particular are "leaky" so that a certain minimum number are required for growth. The protoplasts of

Table 1
EFFECT OF CELL DENSITY
IN FEEDER LAYER ON
PLATING EFFICIENCY OF
MAIZE CELLS.

Concentration of feeder cells		Plating efficiency (%)
Fresh wt (mg/mℓ)	Equivalent to (cells/mℓ)	
	0	1 ± 0.1
0.1	1 × 10³	1 ± 0.1
1.0	1 × 10⁴	10 ± 1.0
5.0	5 × 10⁴	27 ± 1.0
10	1 × 10⁵	42 ± 2.0
20	2 × 10⁵	44 ± 9.0
50	5 × 10⁵	23 ± 2.0
500	5 × 10⁶	2 ± 1.0

From Smith, J. A., Green, C. E., and Gengenbach, B. C., *Plant Sci. Lett.*, 36, 67, 1984. With permission.

Lycopersicon spp. could divide without hormones if protoplasts transformed from *Agrobacterium* were present as feeder system.[158]

XII. SINGLE CELL CULTURE

Culture of an individual cell has also been accomplished. The first successful attempt was to culture single cell of *Nicotiana hybrid* (*N. tabacum* × *N. glutinosa*) in a microchamber[67] employing a conditioned medium.

In this method, a drop of medium along with a single cell is placed on a slide. Following this a drop of mineral oil is placed on either side of a nutrient medium drop and a cover glass is placed on each of these mineral drops. A third cover glass is placed on the nutrient medium bridging the two cover glasses and forming a microchamber. The oil prevents the drying up of the medium and permits gaseous exchange. The whole slide is placed in a petri dish. When the cell colony is sufficiently large it can be removed for experimentation. An advantage of a microchamber is that the cell can be examined in the process of division.

In a follow up, individual cells of tobacco[156] were shown to form colonies in microchambers on a synthetic medium comprised of minerals, sucrose, vitamin, calcium pantothenate, and coconut milk. Thus it was possible to demonstrate that plants are possible from a single cell. Similarly, a single cell of carrot formed a tissue on agar medium enriched with auxin and high nitrogen level, 60mM. This tissue in turn formed embryos.[5]

In a microchamber of the type described above it is possible to culture only one cell at a time, thus precluding its utility for studies on physiology of individual cells, particularly when a number of replicates are required for statistical analysis. The chamber is good enough to demonstrate the totipotency of an individual cell. To overcome this limitation of a microchamber, in an improved version it has been possible to grow individually a large number of cells in small droplets of medium.[77]

The modified microchamber [77](Figure 16) comprises a cover glass (22 × 40 cm) on top of which is placed a cellulose nitrate paper (15 × 15 mm). On the paper is spread a thin layer of mineral oil (0.5 mℓ). In turn on paper are placed microdroplets of culture medium.

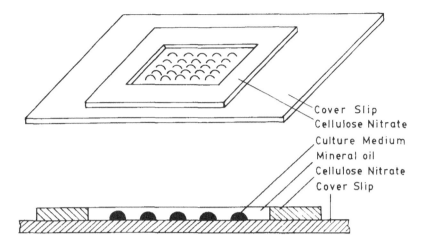

Cover Slip
Cellulose Nitrate
Culture Medium
Mineral oil
Cellulose Nitrate
Cover Slip

FIGURE 16.   Diagrammatic representation of a microchamber for culture of individual cells. (From Koop, H.-U., Weber, G., and Schweiger, H. G., *Z. Pflanzenphysiol.*, 112, 21, 1983. With permission.)

About 100 droplets of 10 $\mu\ell$ or 50 droplets of 25 $\mu\ell$ could be placed on an area of 1 cm². The placing of medium droplets and cells is done employing a micromanipulator. The entire cover glass is placed on the lid of a petri dish and this is in turn placed inside a bigger petri dish (6 cm diameter) having 3 mℓ of distilled water. On culture in this way more than 50% of the individual cells of *Datura innoxia* survived for more than 2 weeks. Of these more than 30% of the cells divided and formed callus tissue. Buffering of medium was an essential requirement for culture of individual cells. Almost all cells died within a few days when unbuffered medium was employed. Addition of MES buffer at 0.1 to 1 mM was found to be essential. Also pH of medium was of critical significance for culture of individual cells. The most favorable pH ranged from 5 to 5.5. Also for cell suspension cultures of *Dioscorea* the pH of the medium had a significant effect. On culture of cells at pH 3.5 to 6.15 the pH of the medium shifted to 4.6 to 4.7 within 10 to 48 hr.[24] Therefore, pH is to be reckoned as an important factor controlling cell growth.

REFERENCES

1. **Aitchison, P. A., Macleod, A. J., and Yeoman, M. M.,** Growth patterns in tissue (callus) cultures, in *Plant Tissue and Cell Culture*, Street, H. E., Ed., Blackwell, Oxford, 1977, 267.
2. **Amino, S., Fujimura, T., and Komamine, A.,** Synchrony induced by double phosphate starvation in a suspension culture of *Catharanthus roseus, Physiol. Plant.*, 58, 393, 1983.
3. **Armstrong, D. J., Murai, N., Taller, B. J., and Skoog, F.,** Incorporation of cytokinin N⁶-benzyladenine into tobacco callus transfer ribonucleic acid and ribosomal ribonucleic acid preparations, *Plant Physiol.*, 57, 15, 1976.
4. **Ashmore, S. and Gould, A. R.,** Karyotype evolution in a tumour-derived plant tissue culture analysed by giemsa C-banding, *Protoplasma*, 156, 297, 1981.
5. **Backs-Husemann, D. and Reinert, J.,** Embryobildung durch isolierte Einzelzellen aus Gewebekulturen von *Daucus carota, Protoplasma*, 70, 49, 1970.
6. **Bagshaw, V.,** Changes in Ultrastructure During the Development of Callus Cells, Ph.D. thesis, University of Edinburgh, Edinburgh, Scotland, 1969.
7. **Balague, C., Latche, A., Fallot, J., and Pech, J. C.,** Some physiological changes occurring during the senescence of auxin-deprived pearl cells in culture. *Plant Physiol.*, 69, 1339, 1982.
8. **Bayliss, M. W.,** Origin of chromosome number variation in cultured plant cells, *Nature (London)*, 246, 529, 1973.

9. **Bayliss, M. W.,** Factors affecting the frequency of tetraploid cells in a predominantly diploid suspension culture of *Daucus carota, Protoplasma,* 92, 109, 1977.

10. **Benbadis, A.,** Croissance de cellules isolees de Rouce sur les milieux mitritifs 'conditionnes' per quantites variables de tissus, *C. R. Acad. Sci.,* 261, 4829, 1965.

11. **Bennici, A., Buiatti, M., D'Amato, F., and Pagliai, M.,** Nuclear behaviour in *Haplopappus gracilis* callus grown in vitro on different culture media, in *Les Cultures de Tissus de Plantes (Colloq. Int. C.N.R.S.),* No. 193, 1971, 245.

12. **Bergmann, L.,** Growth and division of single cells of higher plants in vitro, *J. Gen. Physiol.,* 43, 841, 1960.

13. **Bergmann, L.,** Plating of plant cells, in *Plant Tissue Culture and its Biotechnological Application,* Barz, W., Reinhard, E., and Zenk, M. H., Eds., Springer-Verlag, Berlin, 1977, 213.

14. **Bevan, M. and Northcote, D. H.,** Some rapid effects of synthetic auxins on mRNA levels in cultured plant cells, *Planta,* 152, 32, 1981.

15. **Bharal, S. and Rashid, A.,** Growth of free-cell suspension and plantlet regeneration in the legume *Indigofera enneaphylla, Biol. Plant.,* 26, 202, 1984.

16. **Blakeley, L. M. and Steward, F. C.,** Growth induction in cultures of *Haplopappus gracilis.* I. The behaviour of cultured cells, *Am. J. Bot.,* 48, 351, 1961.

17. **Blakeley, L. M. and Steward, F. C.,** Growth and organized development of cultured cells. V. The growth of colonies from free cells in nutrient agar, *Am. J. Bot.,* 51, 780, 1964.

18. **Blanarikova, V. and Karacsonyi, S.,** The isolation of tissue culture of *Populus alba, Biol. Plant.,* 20, 14, 1978.

19. **Blaschke, J. P., Forche, E., and Neumann, K. H.,** Investigations on cell cycle of haploid and diploid tissue cultures of *Datura innoxia* and its synchronization, *Planta,* 144, 7, 1978.

20. **Breuling, M., Alfermann, A. W. and Reinhard, E.,** Cultivation of cell cultures of *Berberis wilsonae* in 20 L airlift bioreactor, *Plant Cell Rep.,* 4, 220, 1985.

21. **Brossard-Chriqui, D. and Iskander, S.,** Ultrastructural changes of cytomembranes during the first hours of the in vitro culture in *Datura innoxia* leaf explants. Relation to sucrose uptake, *Protoplasma,* 112, 217, 1982.

22. **Browers, M. A. and Orton, T. J.,** A factorial study of chromosomal variablility in callus cultures of celery *(Apium graveolens), Plant Sci. Lett.,* 26, 65, 1982.

23. **Brown, D. J., Canvin, D. T., and Zilkey, B. F.,** Growth and metabolism of *Ricinus communis* endosperm in tissue culture, *Can. J. Bot.,* 48, 2323, 1970.

24. **Butenko, R. G., Lipsky, A. Kh., Chernyak, N. D., and Arya, H. C.,** Changes in culture medium pH by cell suspension cultures of *Dioscorea deltoides, Plant Sci. Lett.,* 35, 207, 1984.

25. **Chibbar, R. N., Cella, R., Albani, D., and Van Huystee, R. B.,** The growth of and peroxidase synthesis by two carrot cell lines, *J. Exp. Bot.,* 35, 1846, 1984.

26. **Chlyah, H., Zair, I., and Chlyah, A.,** In vitro induction of cell division in the epidermis of stem segments of *Torenia fournieri.* Role of subepidermal tissues and amino compounds, *Planta,* 135, 516, 1982.

27. **Cionini, P. G., Bennici, A., and D'Amato, F.,** Nuclear cytology of callus induction and development in vitro. I. Callus from *Vicia faba* cotyledons, *Protoplasma,* 96, 101, 1978.

28. **Conner, A. J. and Meredith, C. P.,** An improved polyurethane support system for monitoring growth in plant cell cultures, *Plant Cell Tissue Organ Culture,* 3, 59, 1984.

29. **Constabel, F., Kurz, W. G. W., Chatson, B., and Gamborg, O. L.,** Induction of partial synchrony in soybean cell cultures, *Exp. Cell Res.,* 85, 105, 1974.

30. **Constabel, F., Kurz, W. G. W., Chatson, B., and Kirkpatrick, J. W.,** Partial synchrony in soybean cell suspension cultures induced by ethylene, *Exp. Cell Res.,* 105, 263, 1977.

31. **Cordon, H., Latche, A., Pech, J. C., Nebie, B., and Fallot, J.,** Control of quiescence and viability in auxin-deprived pear cells in batch and continuous culture, *Plant Sci. Lett.,* 17, 29, 1979.

32. **Dalton, C. C.,** The culture of plant cells in fermenters, Heliosynthese et Aquaculture Seminar de Martiques, CNRS, France, 1978, 1.

33. **D'Amato, F.,** Endopolyploidy as a factor in plant tissue development, *Caryologia,* 17, 41, 1964.

34. **D'Amato, F.,** Cytogenetics of differentiation in tissue and cell cultures, in, *Applied and Fundamental Aspects of Plant Cell Tissue and Organ Culture,* Reinert, J. and Bajaj, Y. P. S., Eds., Springer-Verlag, Berlin, 1977, 343.

35. **De Mey, J., Lambert, A. M., Bajer, A. S., Moeremans, M., and De Brabander, M.,** Visualization of microtubules in interphase and mitotic plant cells of *Hemanthus* endosperm with the immunogold staining method, *Proc. Natl. Acad. Sci. U.S.A.,* 79, 1898, 1982.

36. **Earle, E. D. and Torrey, J. G.,** Colony formaiton by isolated *Convolvulus* cells plated on defined media, *Plant Physiol.,* 40, 520, 1965.

37. **Engvild, K. G.,** Growth and plating of cell suspension cultures of *Datura innoxia, Physiol. Plant.,* 32, 390, 1974.

38. **Eriksson, T.**, Studies on the growth requirements and growth measurements of cell cultures of *Haplopappus gracilis, Physiol. Plant.*, 18, 976, 1965.

39. **Eriksson, T.**, Partial synchronization of cell division in suspension cultures of *Haplopappus gracilis, Physiol. Plant.*, 19, 900, 1966.

40. **Evans, D. A. and Gamborg, O. L.**, Chromosome stability of cell suspension cultures of *Nicotiana* spp., *Plant Cell Rep.*, 1, 104, 1982.

41. **Evans, D. A. and Reed, S. M.**, Cytogenetic techniques, in *Plant Tissue Culture Methods and Application in Agriculture*, Thorpe, T. A., Ed., Academic Press, New York, 1981, 214.

42. **Evans, D. A., Sharp, W. R., and Medina-Filho, H. P.**, Somaclonal and gametoclonal variation, *Am. J. Bot.*, 71, 759, 1984.

43. **Everett, N. P., Wang, T. L., Gould A. R., and Street, H. E.**, Studies on the control of the cell cycle in cultured plant cells. II. Effects of 2,4-dichlorophenoxyacetic acid (2,4-D), *Protoplasma*, 106, 15, 1981.

44. **Favali, M. A., Serafini-Fracassini, D., and Sartorato, P.**, Ultrastructure and autoradiography of dormant and activated parenchyma of *Helianthus tuberosus, Protoplasma*, 123, 192, 1984.

45. **Fowke, L. C., Bech-Hanson, C. W. Constabel, F., and Gamborg, O. L.**, A comparative study on the ultrastructure of cultured cells and protoplasts of soybean during cell division, *Protoplasma*, 81, 189, 1974.

46. **Garcia-Rodriques, M. J.**, Effect of napthaleneacetic acid on histological phenomenon manifested by Jerusalem Artichoke tissue (*Helianthus tuberosuis* L.) variety Blanc Sutton cultured in vitro, *C.R. Acad. Sci. Ser. D.*, 283, 765, 1976.

47. **Gathercole, R. W. E., Mansfield, K. J., and Street, H. E.**, Carbon dioxide as an essential requirement for cultured sycamore cells, *Physiol. Plant*, 37, 213, 1976.

48. **Gautheret, R. J.**, Culture du tissu cambial, *C. R. Acad. Sci.*, 198, 2195, 1934.

49. **Gautheret, R. J.**, Sur la possibilite de realiser la culture indefinie des tissu de tubercules de carotte, *C. R. Acad. Sci.*, 208, 118, 1939.

50. **Gibbs, J. L. and Dougall, D. K.**, Growth of single plant cells, *Science*, 141, 1059, 1963.

51. **Gould, A. R., Bayliss, M. W., and Street, H. E.**, Studies on the growth in culture of plant cells. XVII. Analysis of the cell cycle of asynchronously dividing *Acer pseudoplatanus* cells in suspension culture, *J. Exp. Bot.*, 25, 468, 1974.

52. **Gould, A. R. and Street, H. E.**, Kinetic aspects of synchrony in suspension cultures of *Acer pseudoplatanus, J. Cell Sci.*, 17, 337, 1975.

53. **Gould, A. R., Everett, N. P., Wang, T. L., and Street, H. E.**, Studies on the control of cell cycle in cultured plant cells. I. Effects of nutrient limitation and nutrient starvation, *Protoplasma*, 106, 1, 1981.

54. **Grant, M. E. and Fuller, K. W.**, Tissue culture of root cells of *Vicia faba, J. Exp. Bot.*, 19, 667, 1968.

55. **Guerri, J., Culianez, F., Primo-Millo, E., and Primo-Yufera, E.**, Chromatin changes related to dedifferentiation and differentiation in tobacco tissue culture *(Nicotiana tabacum), Planta*, 155, 273, 1982.

56. **Guilfoyle, T. J., Lin, C. V., Chen, V. M., Nagao, R. T., and Key, J. L.**, Enhancement of soybean RNA polymerase I by auxin, *Proc. Natl. Acad. Sci. U.S.A.*, 72, 69, 1975.

57. **Gunning, B. E. S. and Hardham, A. R.**, Microtubules, *Ann. Rev. Plant Physiol.*, 33, 651, 1982.

58. **Guo, C. L.**, *In Vitro*, 7, 381, 1972.

59. **Guri, A., Zelcher, A., and Izher, S.**, Induction of high mitotic index in *Petunia* suspension cultures by sequential treatment with aphidicolin and colchicine, *Plant Cell Rep.*, 3, 219, 1984.

60. **Haberlandt, G.**, Kultinversuche mit isolierten Pflanzenellen, *Sber. Akad. Wiss. Wien*, 111, 69, 1902.

61. **Hannig, E.**, Zur Physiologie Pflanzlicher Embryonen. I. Uber die kultur von Crufieren-Embryonen ausserhalb des Embryosacks, *Bot Ztg.*, 62, 45, 1904.

62. **Hooker, M. P. and Nabors, M. W.**, Callus initiation, growth and organogenesis in sugarbeet *(Beta vulgaris), Z. Pflanzenphysiol.*, 84, 237, 1977.

63. **Horsch, R. B. and Jones, G. E.**, A double filler paper technique for plating cultured plant cells, *In Vitro*, 10, 102, 1980.

64. **Horsch, R. B., King, J., and Jones, G. E.**, Measurement of cultured plant cell growth on filter paper discs, *Can. J. Bot.*, 58, 2402, 1980.

65. **Jha, T. B. and Roy, S. C.**, Effect of different hormones on *Vigna* tissue culture and its chromosomal behaviour, *Plant Sci. Lett.*, 24, 219, 1982.

66. **Johri, B. M. and Nag, K. K.**, Cytology and morphogenesis of embryo and endosperm tissues in *Dendropthoe* and *Taxillus, Cytologia*, 39, 801, 1974.

67. **Jones, L. E., Hidebrandt, A. C., Ricker, A. J., and Wu, J. H.**, Growth of somatic tobacco cells in microtubules, *Am. J. Bot.*, 47, 468, 1960.

68. **Jouanneau, J. P.**, Controlle par les cytokinines de la synchronisation des mitoses dans les cellules de tabac, *Exp. Cell Res.*, 67, 329, 1971.

69. **Jouanneau, J. P. and Teyssandier, B.**, Synchronisation partielle des mitoses dans des suspensions de cellules de Tabac par addition successive d' auxine et de cytokinine, *C. R. Acad. Sci. Ser. D*, 270, 320, 1970.

70. **Kallack, H. and Yarvekylg, L.,** On the cytogenetic effects of 2,4-D on pea callus in culture, *Acta Biol. Hung.,* 22, 67, 1971.
71. **Kao, K. N. and Michayluk, M. R.,** Nutrient requirements for growth of *Vicia hajastana* cells and protoplasts at a very low population density in liquid media, *Planta,* 126, 105, 1975.
72. **Kao, K. N., Miller, R. A., Gamborg, O. L., and Harvey, B. L.,** Variation in chromosome number and structure in wheat *(Triticum), Can. J. Genet. Cytol.,* 12, 297, 1970.
73. **Katterman, F. R. H., Williams, M. D., and Clay, W. F.,** The influence of a strong reducing agent upon the initiation of callus from the germinating seedlings of *Gossypium barbadense, Physiol. Plant.,* 40, 98, 1977.
74. **Key, J. L.,** Hormones and nucleic acid metabolism, *Ann. Rev. Plant Physiol.,* 20, 449, 1969.
75. **King, P. J., Cox, B. J., Fowler, M. W., and Street, H. E.,** Metabolic events in synchronized cell cultures of *Acer pseudoplatanus, Planta,* 117, 109, 1974.
76. **Komamine, A., Morigaki, T., and Fujimura, T.,** in, *Frontiers of Plant Tissue Culture,* Thorpe, T. A., Ed., University of Calgery Press, Calgery, Alberta, 1978, 159.
77. **Koop, H. U., Weber, G., and Schweiger, H. G.,** Individual culture of single cells and protoplasts of higher plants in microdroplets of defined media, *Z. Pflanzenphysiol.,* 112, 21, 1983.
78. **Kordan, H. A.,** Proliferation of excised juice vesicles of lemon in vitro, *Science,* 129, 779, 1959.
79. **Kordan, H. A.,** The use of mature lemon fruits for tissue culture and other physiological and morphogenetic investigations, *Ann. Bot.,* 53, 137, 1984.
80. **Kumar, P. P., Raju, C. R., Chandramohan, M., and Iyer, R. D.,** Induction and maintenance of friable callus from cellular endosperm of *Cocos nucifera, Plant Sci.,* 40, 203, 1985.
81. **Kurz, W. G. W.,** A chemostat for growing higher plant cells in single cell suspension culture, *Exp. Cell Res.,* 64, 476, 1971.
82. **Lamport, D. T. A.,** Cell suspension cultures of higher plants: isolation and growth energetics, *Exp. Cell Res.,* 33, 195, 1964.
83. **Latche, A. and Pech, J. C.,** Differential capacities of apple and pear fruit explants to enter cell division in vitro during ripening and senescence, *Physiol. Veg.,* 21, 77, 1983.
84. **Leguay, J. J. and Guern, J.,** Quantitative effects of 2,4-dichlorophenoxyacetic acid on growth of suspension-cultured *Acer pseudoplatanus* cells, *Plant Physiol.,* 56, 356, 1975.
85. **Logemann, B. and Bergmann, L.,** Einfluß von Licht und Medium auf die "Plating efficiency" isolierter Zellen aus Kalluskulturen von *Nicotiana tabacum* var. "Samsun", *Planta,* 121, 283, 1974.
86. **Macleod, A. J., Mills, E. D., and Yeoman, M. M.,** Seasonal variations in the pattern of RNA metabolism of tuber tissue in response to excision and culture, *Protoplasma,* 98, 343, 1979.
87. **Mandels, M.,** The culture of plant cells, *Adv. Biochem. Eng.,* 2, 201, 1972.
88. **Matthews, B. F.,** Isolation of mitotic chromosomes from partially synchronized carrot cell suspension cultures, *Plant Sci. Lett.,* 31, 165, 1983.
89. **Mehta, A. R.,** Recent advances in free cell cultures of plants, in *Tissue Culture,* Ramakrishnan, C. V., Ed., Dr. W. Junk Publishers, The Hague, 1965, 305.
90. **Melanson, D. L. and Ingle, J.,** Regulation of ribosomal RNA accumulation by auxin in artichoke tissue, *Plant Physiol.,* 62, 761, 1978.
91. **Melchers, G. and Bergmann, L.,** Untersuchungen an kulturen von haploid Geweben von *Antirrhinum majus, Ber Dtsch. Bot. Ges.,* 71, 459, 1959.
92. **Melchers, G. and Engelmann, D.,** Die kultur von Pflanzengewebe ein flussigen medium mit Dauerbeluftung, *Naturwissenschaften,* 42, 564, 1955.
93. **Mii, M., Saxena, P. K., Fowke, L. C., and King, J.,** Isolation of chromosomes from cell suspension cultures of *Vicia hajasthana, Cytologia,* in press.
94. **Miller, R. A., Shyluk, J. P., Gamborg, O. L., and Kirkpatrick, J. W.,** Phytostat for continuous culture and automatic sampling of plant cell suspensions, *Science,* 159, 540, 1968.
95. **Mitra, J. and Steward, F. C.,** Growth induction in cultures of *Haplopappus gracilis.* III. The behaviour of the nucleus, *Am. J. Bot.,* 48, 358, 1961.
96. **Moloney, M. M., Hall, J. F., Robinson, G. M., and Elliot, M. C.,** Auxin requirements of sycamore cells in suspension culture, *Plant Physiol.,* 71, 927, 1983.
97. **Monod, J.,** *Ann. Inst. Pasteur Paris,* 79, 390, 1950.
98. **Muhitch, M. J., Edward, L. A., and Fletcher, J. S.,** Influence of diamines and polyamines on the senescence of plant suspension cultures, *Plant Cell Rep.,* 2, 82, 1983.
99. **Muir, W. H., Hildebrandt, A. C., and Riker, A. J.,** Plant tissue cultures produced from single isolated cells, *Science,* 119, 877, 1954.
100. **Nagata, T., Okada, K., and Takebe, I.,** Mitotic protoplasts and their infection with tobacco mosaic virus RNA encapsulated in liposomes, *Plant Cell Rep.,* 1, 250, 1982.
101. **Nakagawa, K., Fukui, H., and Tabata, M.,** Hormonal regulation of berberine production in cell suspension cultures of *Thalictrum minus, Plant Cell Rep.,* 5, 69, 1986.

102. **Nandi, S., Fridborg, G., and Eriksson, T.,** Effects of 6-(3-methyl-2-buten-1-ylamino)-purine and α-naphthaleneacetic acid on root formation and cytology of root tips and callus tissues of *Allium cepa* var. prolifera, *Hereditas,* 85, 57, 1977.

103. **Nash, D. T. and Davies, M. E.,** Some aspects of growth and metabolism of Paul's scarlet rose cell suspensions, *J. Exp. Bot.,* 23, 75, 1972.

104. **Neskovic, M., Vujicic, R., and Srejovic, V.,** Differential responses of buck wheat leaf cells to growth substances stimulating cell division, *Ann. Bot.,* 56, 755, 1985.

105. **Nishi, A., Kato, K., Takahashi, M., and Yoshida, R.,** Partial synchronization of carrot cell culture by auxin deprivation, *Physiol. Plant.,* 39, 9, 1977.

106. **Nishinari, N. and Syono, K.,** Induction of cell division synchrony and variation in cytokinin contents through the cell cycle in tobacco cultured cells, *Plant Cell Physiol.,* 27, 147, 1986.

107. **Nobecourt, P.,** Sur la perennite et l'augmentation de volume des cultures de tissus vegetaux, *C. R. Seanc. Soc. Biol.,* 130, 1270, 1939.

108. **Nobecourt, P.,** Sur les radicelles naissant des cultures de tissus vegetaux, *C. R. Seanc. Soc. Biol.,* 130, 1271, 1939.

109. **Okamura, S., Mijasaka, K., and Nishi, A.,** Synchronisation of carrot cell culture by starvation and cold treatment, *Exp. Cell Res.,* 78, 467, 1973.

110. **Park, K.-H., Fujisawa, S., Sakurai, A., Yamaguchi, I., and Takahashi, N.,** Gibberellin production in cultured cells of *Nicotiana tabacum, Plant Cell Physiol.,* 24, 1241, 1983.

111. **Park, K.-H., Fujisawa, N., Sakurai, A., Yamaguchi, I., and Takahashi, N.,** Changes in endogenous gibberellin contents during growth of cultured tobacco cells, *Plant Cell Physiol.,* 25, 1303, 1984.

112. **Partanen, C. R.,** Quantitative chromosomal changes and differentiation in plants, in *Developmental Cytology,* Rudnick, D., Ed., Ronald Press, New York, 1959, 21.

113. **Puschmann, R., Ke, D., and Romani, R.,** Ethylene production by suspension-cultured pear fruit cells as related to their senescence, *Plant Physiol.,* 79, 973, 1985.

114. **Puschmann, R. and Romani, R.,** Ethylene production by auxin deprived, suspension-cultured pear fruit cells in response to auxins, stress or precursor, *Plant Physiol.,* 73, 1013, 1983.

115. **Ramaswamy, O., Pal, S., Guha-Mukerjee, S., and Sopory, S. K.,** Correlation of glyoxalase I activity with cell proliferation in *Datura* callus culture, *Plant Cell Rep.,* 3, 121, 1984.

116. **Rashid, A. and Street, H. E.,** Growth, embryogenic potential and stability of haploid cell culture of *Atropa belladonna, Plant Sci. Lett.,* 2, 89, 1974.

117. **Reinert, J.,** Untersuchungen uber die morphogenese an Gewebekulturen, *Ber. Dtsch. Bot. Ges.,* 71, 15, 1958.

118. **Reinert, J.,** Growth of single cells from higher plants on synthetic media, *Nature (London),* 200, 90, 1963.

119. **Reinert, J. and White, P. R.,** The cultivation in vitro of tumour tissues and normal tissues of *Picea glauca, Physiol. Plant.,* 9, 177, 1956.

120. **Romani, R. J., Boss, T. J., and Pech, J. C.,** Cycloheximide stimulation of cyanide-resistant respiration in suspension cultures of senescent pear fruit cells, *Plant Physiol.,* 68, 823, 1981.

121. **Röper, W., Schultz, M., Chaouiche, E., and Meloh, K. A.,** Nicotine production by tissue cultures of tobacco as influenced by various culture parameters, *J. Plant Physiol.,* 118, 463, 1985.

122. **Royle, D. L. and Cleland, R.,** Control of plant cell enlargement by hydrogen ions, in *Current Topics in Developmental Biology,* Vol. II, Monroy, A., and Moscona, A. A., Eds., Academic Press, New York, 1977.

123. **Sacristan, M. D.,** Karyotypic changes in callus cultures from haploid and diploid plants of *Crepis cepillaris, Chromosoma,* 33, 273, 1971.

124. **Schmidt, A. C., Vantard, M., de Mey, J., and Lembert, A. M.,** Aster-like microtubule centers establish spindle polarity during interphase-mitosis transition in higher plant cells, *Plant Cell Rep.,* 2, 285, 1983.

125. **Sembdner, G., Gross, D., Lebisch, H. W., and Schneider, G.,** Biosynthesis and metabolism of plant hormones, in *Encyclopaedia of Plant Physiology Hormonal Regulation of Development,* Vol. 1, Macmillan, J., Ed., Springer-Verlag, Berlin, 1980, 345.

126. **Shridan, W. F.,** Plant regeneration and chromosome stability in tissue cultures, in *Genetic Manipulations with Plant Material,* Ledou, L., Ed., Plenum Press, New York, 1975, 263.

127. **Short, K. C., Brown, E. G., and Street, H. E.,** Studies on the growth in culture of plant cells. V. Large scale culture of *Acer pseudoplatanus* cell suspension. *J. Exp. Bot.,* 20, 579, 1969.

128. **Simpkins, I., Collin, H. A., and Street, H. E.,** The growth of *Acer pseudoplatanus* cells in a synthetic liquid medium: response to the carbohydrate, nitrogenous and growth hormones constituents, *Physiol. Plant.,* 23, 385, 1970.

129. **Singh, B. D.,** Chemical composition of medium in relation to cytogenetic behaviour of *Haplopappus gracilis* callus cultures, *Nucleus,* 19, 74, 1976.

130. **Singh, B. D. and Harvey, B. L.,** Selection for diploid cells in suspension cultures of *Haplopappus gracilis, Nature (London),* 253, 453, 1975.

131. **Smart, N. J. and Fowler, M. W.,** An airlift column bioreactor suitable for large scale cultivation of plant cell suspension, *J. Exp. Bot.,* 35, 531, 1984.

132. **Smith, J. A., Green, C. E., and Gengenbach, B. G.,** Feeder layer support of low density population of *Zea mays* suspension cells, *Plant Sci. Lett.,* 36, 67, 1984.

133. **Smith, B. A., Reider, M. L., and Fletcher, J. S.,** Relationship between vital staining and subculture growth during the senescence of plant tissue culture, *Plant Physiol.,* 70, 1228, 1982.

134. **Snijman, D. A., Noel, A. R. A., Bornman, C. H., and Abbott, J. G.,** *Nicotiana tabacum* callus studies. II. Variability in cultures, *Z. Pflanzenphysiol.,* 82, 367, 1977.

135. **Srivastava, P. S.,** In vitro induction of triploid roots and shoots from mature endosperm of *Jatropha panduraefolia, Z. Pflanzenphysiol.,* 66, 93, 1971.

136. **Steward, F. C., Caplin, S. M., and Miller, F. K.,** Investigations of growth and metabolism of plant cells. I. New techniques for the investigation of metabolism, nutrition and growth in undifferentiated cells, *Ann. Bot.,* 16, 58, 1952.

137. **Steward, F. C., Mapes, M. O., Kent, A. E., and Holsten, R. D.,** Growth and development of cultured plant cells, *Science,* 163, 20, 1964.

138. **Steward, F. C. and Shantz, E. M.,** The chemical induction of growth in plant tissue cultures, in *The Chemistry and Mode of Action of Plant Growth Substances,* Wain, R. L. and Wightman, F., Eds., Butterworths, London, 1965, 165.

139. **Street, H. E.,** Cell suspension cultures — techniques, in *Plant Tissue and Cell Culture,* Street, H. E., Ed., Blackwell, Oxford, 1977, 61.

140. **Street, H. E., King, P. J., and Mansfield, K.,** Growth control in plant cell suspension cultures, *Les Cultures de Tissus de Plantes,* Colloques Internationaux CNRS, Paris, 193, 17, 1971.

141. **Stuart, R. and Street, H. E.,** Studies on the growth in culture of plant cells. IV. The initiation of division in suspension of stationary phase cells of *Acer pseudoplatanus, J. Exp. Bot.,* 20, 556, 1969.

142. **Stuart, R. and Street, H. E.,** Studies on the growth in culture of plant cells. X. Further studies on the conditioning of culture media by suspension of *Acer pseudoplatanus, J. Exp. Bot.,* 22, 96, 1971.

143. **Sunderland, N.,** Nuclear cytology, in *Plant Tissue and Cell Culture,* Street, H. E., Ed., Blackwell, Oxford, 1977, 177.

144. **Szabados, L., Hadleczky, Gy, and Dudits, D.,** Uptake of isolated plant chromosomes by plant protoplasts, *Planta,* 151, 141, 1981.

145. **Szent-Gyorgyi, A.,** On retine, *Proc. Natl. Acad. Sci. U.S.A.,* 57, 1642, 1967.

146. **Teyssendier, B. and Jouanneau, J. P.,** Preferential incorporation of an exogenous cytokinin N^6-benzyladenine into 18S and 23S ribosomal RNA of tobacco cells in suspension culture, *Biochimie,* 61, 913, 1979.

147. **Teyssendier, B., Jouanneau, J. P., and Peaud-Lenoel, C.,** Covalent linkage of N^6-(Δ-isopentyl) adenine nucleotide to the rRNA of cytokinin requiring and of cytokin-autonomous tobacco cells, *J. Plant Growth Regul.,* 1, 25, 1982.

148. **Teyssendier, B., Jouanemau, J. P., and Peaud-Lenoel, C.,** Incorporation of N^6-benzyladenine into messenger poly(A)-RNA of tobacco cells incubated at stimulatory or cytostatic cytokinin concentration, *Plant Physiol.,* 74, 669, 1984.

149. **Torrey, J. G.,** Cell division in isolated single cells in vitro, *Proc. Natl. Acad. Sci. U.S.A.,* 43, 887, 1957.

150. **Torrey, J. G.,** Cytological evidence of cell selection by plant tissue culture media, in *Int. Conf. Plant Tissue Culture,* White, P. R. and Grove, A. R., Eds., McCutchan Publishing, Berkeley, Calif., 1965, 473.

151. **Torrey, J. G. and Shigemura, J.,** Growth and controlled morphogenesis in pea root callus tissue grown in liquid media, *Am. J. Bot.,* 44, 334, 1957.

152. **Tulecke, W. and Nickell, L. G.,** Methods, problems and results of growing plant cells under submerged condition, *Trans. N.Y. Acad. Sci.,* 22, 196, 1960.

153. **Van Huystee, R. B. and Cairns, W. L.,** Appraisal of studies on induction of peroxidase and associated porphyrin metabolism, *Bot. Rev.,* 46, 429, 1980.

154. **Van Huystee, R. B. and Lobarzewski, J.,** An immunological study of peroxidase release by cultured peanut cells, *Plant Sci. Lett.,* 27, 59, 1982.

155. **VanZulli, L., Magnien, E., and Olivi, L.,** Caryological stability of *Datura innoxia* calli analysed by cytophotometry for 22 hormonal combinations, *Plant Sci. Lett.,* 17, 181, 1980.

156. **Vasil, V. and Hildebrandt, A. C.,** Differentiation of tobacco plants from single isolated cells in microculture, *Science,* 50, 889, 1965.

157. **Veliky, I. A. and Martin, S. M.,** A fermenter for plant cell suspension cultures, *Can. J. Microbiol.,* 16, 223, 1970.

158. **Vidair, C. and Hanson, M. R.,** *Agrobacterium*-transformed tomato cells replace the hormone requirement for growth of tomato leaf protoplasts, *Plant Sci.,* 41, 185, 1985.

159. **Wang, W. C., Beyl, C., and Sharma, G. C.,** A comparison of ward wheat suspension cultures containing clumps and single cells, *Plant Sci. Lett.,* 31, 147, 1983.

160. **Webb, D. T.**, Light-induced callus formation and root growth inhibition of *Dioon edule* seedlings in sterile culture, *Z. Pflanzenphysiol.*, 106, 223, 1982.

161. **Webb, D. T.**, Developmental anatomy and histochemistry of light-induced callus formation by *Dioon edule* (Zamiaceae) seedling roots in vitro, *Am. J. Bot.*, 71, 65, 1984.

162. **Weber, G. and Lark, K. G.**, An efficient plating system for rapid isolation of mutants from plant cell suspensions, *Theor. Appl. Genet.*, 55, 81, 1979.

163. **White, P. R.**, Potentially unlimited growth of excised tomato root tips in a liquid medium, *Plant Physiol.*, 9, 585, 1934.

164. **White, P. R.**, Vitamin B_1 in the nutrition of excised tomato roots, *Plant Physiol.*, 12, 803, 1937.

165. **White, P. R.**, Potentially unlimited growth of excised plant callus in an artificial medium, *Am. J. Bot.*, 26, 59, 1939.

166. **Wilson, G. and Marron, P.**, Growth and anthraquinone biosynthesis by *Galium mollugo* cells in batch and chemostat cultures, *J. Exp. Bot.*, 29, 837, 1978.

167. **Wilson, S. B., King, P. J., and Street, H. E.**, Studies on the growth in culture of plant cells. XII. A versatile system for the large scale batch or continuous culture of plant cell suspensions, *J. Exp. Bot.*, 21, 177, 1971.

168. **Wilson, H. M. and Street, H. E.**, The growth, anatomy and morphogenetic potential of callus and cell suspension cultures of *Hevea brasiliensis*, *Ann. Bot.*, 39, 671, 1975.

169. **Wink, M.**, Composition of the spent cell culture medium. I. Time course of ethanol formation and the excretion of hydrolytic enzymes into the medium of suspension-cultured cells of *Lupinus polyphyllus*, *J. Plant Physiol.*, 121, 287, 1985.

170. **Winkler, H.**, Uber Merogonie und Befruchtung, *Jahrb. Wiss. Bot.*, 36, 753, 1901.

171. **Yeoman, M. M.**, Early development in callus cultures, *Int. Rev. Cytol.*, 29, 383, 1970.

172. **Yeoman, M. M. and Davidson, A. W.**, Effect of light on cell division in developing callus cultures, *Ann. Bot.*, 35, 1085, 1971.

173. **Yeoman, M. M., Dyer, A. F., and Robertson, A. I.**, Growth and differentiation of plant tissue cultures. I. Changes accompanying the growth of explants from *Helianthus tuberosus* tubers, *Ann. Bot.*, 29, 265, 1965.

174. **Yeoman, M. M. and Evans, P. K.**, Growth and differentiation of plant tissue cultures. II. Synchronous cell divisions in developing callus cultures, *Ann. Bot.*, 31, 323, 1967.

175. **Zeleneva, I. V. and Khavkin, E. E.**, Rearrangement of enzyme patterns in maize callus and suspension cultures, *Planta*, 148, 108, 1980.

Chapter 2

CELL DIFFERENTIATION

I. INTRODUCTION

Differentiation of a cell has been of considerable interest, but its understanding is far from complete. The significance of cell differentiation lies in the fact that during the development of a multicellular organism, information from a variety of sources, genetic and otherwise, interacts and influences the expression of certain cells so that they become specialized to perform particular functions. For an understanding of cell specialization it is essential to have an identification of these signals and a determination of their roles. However, it is difficult to follow different functions in a multicellular organism. By contrast in cell cultures the signals cannot only be easily defined and programed, but their interactions also followed.

Differentiation is an all-inclusive word. In it one can include changes occurring at the level of physiology and biochemical potential of a cell which may be with or without a morphological manifestation. When expressed at a morphological level, the differentiation of a cell may be limited to the level of an organelle or it can encompass an entire cell. In this account are taken some representative examples of cell differentiation and their controls defined.

II. PHYSIOLOGICAL DIFFERENTIATION — CELL HABITUATION

Plant cells on prolonged culture, either on agar or liquid medium, reveal changes in their physiological potentialities, particularly in respect of hormone requirements. At times, they become hormone autonomous and this change has been described as cell habituation.

A. Auxin Habituation

The first recorded physiological change was in respect of auxin requirement. Tissue of *Scorzonera hispanica* on prolonged culture became independent of auxin — auxin autotrophic.[56] Other auxin-independent cell lines known are from grape,[108] tobacco,[45] crepis,[134] sunflower and Vinca,[143] sycamore,[38,88] and sugarbeet.[32] Also cell lines of carrot resistant to 5-methyl-tryptophan (5-MT) were found to be auxin autotrophic.[170] Since tryptophan is a precursor of auxin, IAA, it is not surprising that providing it from outside will eliminate the requirement for exogenous auxin. In fact, a cell line of carrot[145] resistant to 5-MT contained an increased amount of IAA. However, none of the tobacco cell lines resistant to 5-MT were auxin autotrophic.[171]

Very little is known about the mechanism of auxin autonomy. In auxin-autotrophic tissue of sugarbeet it has been demonstrated, employing very sensitive techniques, that although there is no difference between endogenous content of IAA in two types of tissues, auxin-requiring normal tissue contains higher levels of IAA-oxidase than habituated tissue.[72] This lends credence to an earlier suggestion, which was considered to be controversial at that time, that there is an increased rate of enzymatic inactivation of auxin in auxin-requiring tissue.[117,146,147] Also, turnover of free IAA is higher in autotroph than heterotroph callus[78] and there is less bound IAA in autotroph than heterotroph tobacco callus.

B. Cytokinin Habituation

Habituation has also been reported in respect to cytokinin requirement and is of more frequent occurrence than auxin habituation. Cytokinin habituation has been reported by a number of workers in tobacco tissue.[46,68,99,164] Other tissues which show cytokinin habituation are citrus[24] and blackberry.[54,55]

A property of cytokinin autonomous tobacco cells described is their tolerance to toxic levels of bromodeoxyuridine (Budr). This is in common to crown gall cells.[11,12,163] However, this aspect is to be looked into in greater detail because cytokinin habituation and Budr resistance are independent characters in tobacco.[68]

Only in tobacco tissue[45] have both auxin autotrophy as well as cytokinin autotrophy been recorded. The two contrasting viewpoints about the relationship between auxin and cytokinin autotrophy are an apparent reciprocity[37] to no linkage.[164] Cytokinin activity is a property of two distinct types of compounds, N^6-substituted adenine derivatives and substituted phenylureas. The tissue of *Phaseolus lunatus* cv. Jackson wonder became cytokinin autonomous on any concentration of thidiazuron (a phenylurea compound), but tissues from other genotypes remained cytokinin dependent.[23]

1. Induction of Cytokinin Habituation

It has been possible to induce cytokinin habituation in tobacco tissue. Explants from parenchyma tissue of Havana 425 cultivar of tobacco on incubation at 35°C for 7 days on an auxin-containing medium formed hyperplastic nodules which when separated from the tissue were capable of growth on auxin medium alone, and were cytokinin autotrophic.[102] The incidence of habituation was promoted by cytokinin, at 1000-fold lower concentration than required for growth of normal tissue. Also, explants taken during spring habituated seven times faster than tissue isolated from plants during winter. Further, it was found that there is longitudinal gradient in cytokinin habituation frequency.[103] Explants from base habituated infrequently whereas cells from less mature pith habituated more frequently.

In another cultivar of tobacco, Maryland Mammoth or Wisconsin 38, habituation was also possible at ordinary temperature (25°C); high temperature was not required.[160] Explants from the apical region, 1 cm from apex, habituated at a frequency of 100%, within 5 days, contrary to this, explants from 40 cm or more from apex habituated at 18% frequency only and took 2 weeks to grow. Also, in an explant from mature pith the peripheral tissue was more prone to habituation than tissue from pith center.

An analysis of cytokinin habituation indicates that it is an epigenetic[100] rather than a genetic change. This is because the phenomenon is gradual, directed, and reversible, and it leaves the tissue totipotent and occurs in frequencies greatly in excess of those expected for mutation. Furthermore, in order to explain habituation, it has been proposed that either there is a positive feedback loop in which cytokinin either induces its own synthesis or inhibits its own degradation.

Cells from tobacco pith are described to be in a state of determination in respect of cytokinin habituation. There are cells which are inducible and habituate readily when treated with kinetin or incubated at 35°C. And there are noninducible cells which do not habituate under these conditons.[101] This is in contrast to cells from leaf which are exclusively of noninducible type. In addition, habituated cells can be classified into two types, cells which give rise to normal plants whose leaf and pith cells require cytokinin for growth and habituated cells which give rise to plants whose cells do not require cytokinin for growth.

C. Loss of Habituation

Both auxin as well as cytokinin habituations are reversible. The reversion can either be spontaneous or induced. The latter is of more interest. A cytokinin-habituated tissue of tobacco[147] was reversibly lost when cultures were grown at 16°C instead of 25°C. Also, auxin-habituated cultures of *Helianthus annus* and *Vinca rosea*[143] were lost on transfer from White's mineral medium to a mineral medium supplemented with extra levels of KCl, $NaNO_3$, NaH_2PO_4, and $(NH_4)_2SO_4$.

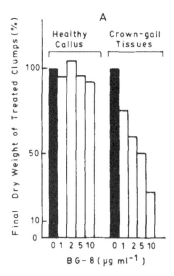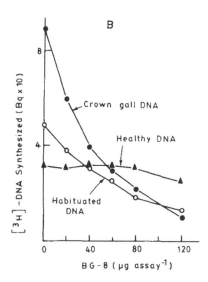

FIGURE 1.   Inhibition of growth (A) and DNA synthesis (B) of crown gall cells by alstonine (BG-8), an anticancer alkaloid. (From LeGoff, L., Roussaux, J., Aaron-da Cunha, M. I., and Beljanski, M., *Physiol. Plant.*, 64, 177, 1985. With permission.)

III. NEOPLASTIC CELL

A neoplastic cell is capable of continuous proliferation on a hormone-free medium. Such cells normally do not occur in higher plants, but can be induced on them. They are the manifestation of either a disease or a genetic imbalance arising out of hybridization. The formation of a gall at the junction of shoot and root (crown region and consequently described as crown gall) incited by *Agrobacterium tumefaciens* represents the former and the presence of genetic tumors on interspecific hybrids such as *N. glauca* × *N. langsdorffii* represent the latter.

The cells from a tumor are not only self-sustaining on the plant they arise, but develop into sizable tumors when grafted onto healthy plants. On isolation from plant, crown gall cells or cells from a genetic tumor are capable of indefinite proliferation in vitro on a hormone-free medium. Neoplastic cells are a parallel plant system for the study of tumor biology. This viewpoint is substantiated by a recent finding of growth inhibition of crown gall cells of *Nicotiana* and *Parthenocissus* by alstonine (BG-8) an alkaloid, the anticancer effect of which has been demonstrated on animal system. The growth of normal cells is not affected. This alkaloid binds to DNA of crown gall cells, thereby inhibiting its synthesis, whereas it has no effect on DNA of normal cells[85] (Figure 1).

A. Genetic Tumor Cell

It is worth recalling that tumor cells of tobacco hybrid[169] (*N. glauca* × *N. langsdorffii*) were among the first cells to be grown in vitro. These cells are capable of indefinite proliferation in vitro on a hormone-free medium.

Formation of tumors is a quite common feature on certain hybrids within the genera *Nicotiana, Brassica, Bryophyllum,* and *Lycopersicon*. The tumors arise at any stage in the development, but are more common on mature plants when vegetative growth comes to an end. Tumors are more frequent on roots, but also arise on aerial parts. Instances of tumor formation are well worked out in tobacco and there are at least 30 to 40 interspecific combinations showing spontaneous tumor formation on hybrid plants.[142] Plants in these combinations can be grouped into + or − category and only when plants of + and −

categories are hybridized, the tumors appear on the offsprings.[115] Further, within each category it is possible to grade plants in their potentiality to form tumors, indicating thereby that tumor formation has a cytogenetic basis.[1,71]

Cells from a tumor as well as from other regions of a tumor-prone hybrid plant are capable of indefinite proliferation on a hormone-free medium. These cells are different from cells of parental species which require auxin and cytokinin for proliferation. The cells from tumor-prone hybrids of *N. glauca* × *N. langsdorffii* and *N. longiflora* × *N. debneyi*[7] had a higher concentration of auxin than the cells from parent species. Also, a higher concentration of scopoletin and scopolin was found in cells from tumor-prone hybrid *N. glauca* × *N. langsdorffii* than in the cells from parent species. These inhibitors of IAA-oxidase possibly regulate level of auxin by reducing its degradation.[157]

B. Crown Gall Cell

Crown gall is a disease symptom of most of the dicotyledonous plants brought about by *Agrobacterium tumefaciens*. This bacterium is a member of Rhizobiaceae; different strains differ in their extent of virulence and can be characterized by the morphology of the induced tumor. Strain B_6 is highly virulent and produces large tumors which remain unorganized, and strain T_{37} forms slow-growing tumors which often turn into teratomas (shows differentiation of shoots), particularly in those plants which have high regenerative capacity such as *Nicotiana* and *Kalanchoe*.

For induction of tumors, stock cultures of *Agrobacterium tumefaciens* are maintained on a nutrient agar (Oxoid CM3 agar 17 g, 1 g yeast extract and 5 g sucrose in a liter of water) and stored at 4°C. Prior to induction of tumors the culture can be transferred to liquid medium and within 10 hr the growth is exponential and bacteria are ready for inoculation.

Induction of tumors is possible on any surface, preferably a stem or even leaf vein. However, a prerequisite is a wounded surface and after 1 to 2 days of wounding when in response to it, cells prepare for proliferation, the cut surface is inoculated with bacteria. To prevent further infection the surface is sealed. After 3 to 5 days the sealing is removed. Within 2 weeks of inoculation tumors can be seen. An ideal temperature for tumor formation is 25°C; a higher temperature, 32°C, is inhibitory.

After initiation of tumors the presence of the bacterium is not essential for proliferation of tumorous cells or for the formation of secondary tumors, at a distance, away from primary tumors. It is also possible to eliminate bacterial cells from tumor cells by subjecting the explant-carrying tumor to high temperature (46 to 47°C) for 5 to 7 days.[16] At this temperature bacteria are selectively killed. To prevent the damage to tumor cells an increase in relative humidity is helpful.

Cells in a crown gall are transformed by bacterium and this transformation is revealed by their capacity of continuous proliferation on a hormone-free medium. These cells are also characterized by the synthesis of unusual amino acid derivatives, opines.[110,121] The presence of a specific opine is correlated with the bacterial strain. In galls induced by strain B_6 of *Agrobacterium,* octopine (*N*-α-[1-carboxyethyl]-L-arginine) is found whereas in galls induced by strain T_{37}, nopaline (*N*-α-glutaryl-L-arginine) is found. These unusual amino acids cannot be metabolized by higher plant cells, but can be catabolized by the *Agrobacterium*. Therefore, it is very likely that a cell transformed by this bacterium resulting in formation of a gall at the soil surface or on roots is a source of amino acid to other agrobacteria living in soil.

IV. DIFFERENTIATION OF BIOSYNTHETIC POTENTIAL

Plant cell cultures, analogous to microbes, are reckoned to be a potential source of useful chemicals such as pharmaceuticals, agrochemicals, food additives, beverages, and cosmetics. This is due to the fact that the supply of these natural plant products is often reduced on the plant due to environmental stresses.

However, unfortunately, cells in culture either do not produce or tend to produce the desired metabolites in much lower quantities. It is possible to select cell lines that produce more substantial amounts of secondary metabolites than normal, but such cell lines are genetically and metabolically unstable and a continuous selection is needed to maintain productivity.[181] Exceptions to this generalization are production of diosgenin,[69,73] Ginseng saponins,[53] hermin,[126] and visnagin[149] in cell cultures at levels approaching or even exceeding those of the parent plant. Added to this list is ubiquinone-10 which is used in Japan for cure of congestive cardiac diseases. A cell line of tobacco has been isolated which accumulates higher levels of ubiquinone-10 than any other plant or microorganism.[95]

One of the reasons for the inability of plant cells to produce a desired metabolite on culture can be their growth in an alien environment. The other reason could be that cells, in a culture, are in a state of differentiation/organization which is different from that found in the intact plant. Possibly, organization/differentiation of cells is an essential feature for normal cell metabolism. This is supported by the finding that root-forming callus of *Atropa belladonna* was capable of producing tropane alkaloids and not an undifferentiated tissue.[150] Similarly, differentiated tissue cultures of *Papaver bracteatum*[67] and *N. tabacum*[119] produced more thebaine and nicotine, respectively, than comparable undifferentiated tissues. This viewpoint is also supported by the observation that suspension culture of *P. somniferum* is most productive in opiate alkaloids when a high number of specialized laticiferous cells are present.[111]

Many of the most important plant compounds are formed or accumulate only in specialized cells. For instance, cardiac glycosides of *Digitalis* are mainly found in leaf cells, quinine and quinidine accumulate in bark of *Cinchona,* tropane alkaloids are synthesized in roots of *Solanaceae* and accumulate in leaves. During plant growth the cells become specialized, morphologically as well as chemically. In cell cultures morphological specialization is repressed and it is very likely that this is also accompanied by loss of chemical specialization. In contrast to it, rapidly growing cell suspension of *Catharanthus roseus,* showing little differentiation, produced serpentine which paralleled the growth.[44] This indicates that biosynthetic potential of plant cells remains latent and is not lost; possibly it needs proper stimulus for expression. Numerous factors have been worked out which affect biosynthetic potential of cells. For an extended reading the readers are referred to recent reviews[44,93,132] on the subject. The main factors which have been observed to affect the synthesis are nutrients, hormone level, irradiance, temperature, pH, aeration, etc. However, so far, the manipulation of culture media, cultural conditions, and hormone levels have, in general, failed to permit commercial production of phytochemicals useful in medicine and industry. A notable exception is the production of shikonin by *Lithospermum erythrorhizon* cells.[47] A similar exception is nicotine production in tobacco callus by the supply of organic acids resulting in production of high level of total alkaloid as compared to leaves of donor plants.[151]

One of the reasons for our failure in exploiting the potential of cell cultures in the production of desired chemicals is a lack of fundamental knowledge of biosynthetic pathways and their associated enzyme systems. This is supported by encouraging results that have been possible due to precursor feeding, use of elicitors, biotransformation, and immobilization of cells.

A. Precursor Feeding

Experiments on precursor feeding should be widely attempted. One of the most successful examples is the increase in quinoline alkaloids in *Ruta graveolens* cultures on addition of 4-hydroxy-2-quinoline.[14] A more recent report is enhancement of anthraquinone synthesis by *Galium* cell cultures, in phosphate-limiting condition, by feeding orthosuccinyl benzoic acid, a precursor of anthraquinone[173] (Figure 2).

B. Use of Elicitor

Microbial insult of normal plants leads to the induction and accumulation of antibiotic

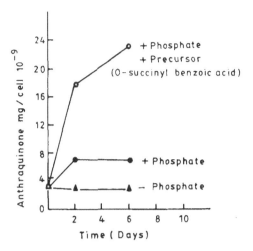

FIGURE 2.   Biosynthesis of anthraquinone in *Galium* cells, following addition of a precursor, orthosuccinyl benzoic acid in phosphate limited condition. (From Wilson, G. and Balague, C., *J. Exp. Bot.*, 36, 485, 1985. With permission.)

biochemicals which are secondary metabolites. The microbial attack or association, viral, bacterial, or fungal, leading to synthesis of antibiotic or other phytochemicals is due to microbe-derived molecules or elicitors, various ones of which have been identified.[13,166] The biotic elicitors include glucan polymers, glycoproteins, low-molecular-weight organic acids, and fungal cell wall material. Abiotic elicitors include UV radiation, salts of heavy metals, and chemicals, some of which have affinity for double-stranded DNA while others disturb cellular membrane integrity.[31,97] Although many molecules act as elicitors, the most extensively used are fungal carbohydrates. Proteinaceous and other organic acid elicitors are also known. In some cases, several active elicitors have been isolated from one source and they show different biological activities. Elicitors can induce a rapid and dramatic increase in the yield of phytochemicals from higher plant cells in suspension culture. An addition of 2-diethyl-aminoethyl-2,4-dichlorophenylether to *Catharanthus roseus* cell cultures resulted in accumulation of indole alkaloids, ajmalicine and cathranthine[86] (Figure 3).

It has been shown that the growing of plant cells in the presence of microorganisms, such as yeast, fungi, and bacteria, results in increased synthesis of secondary metabolites such as alkaloids,[36] an isocoumarin,[80] and sesquiterpenes.[148] In particular, it is to be mentioned that a co-culture of yeast, *Rhodotorula rubra,* with cells of *Ruta graveolens* resulted in a 100-fold increase in synthesis of acridone epoxide alkaloid.[36] A report of recent origin indicates that the synthesis of psoralen in cells of parsley is induced by fungal elicitors.[152] The derivatives of this compound are used both in the treatment of psoriasis and as ingredients of photosensitizing suntan lotions. Similarly, autoclaved fungal mycelia associated with roots of plants in nature increased diosgenin production in *Dioscorea*[131] cell suspension culture and the addition of autoclaved mycelia of *Aspergillus* and *Phytopthora* increased anthraquinone content of *Cinchona* cell cultures.[172]

Although this co-culture approach results in an increased secondary metabolite production, the alterations in metabolism are unpredictable and an empirical screening approach is necessary. Also it is not known how elicitors initiate secondary metabolite production in cultured plant cells. How do elicitors interact with plant cell wall or plasmalemma? Are there elicitor-specific receptor sites? How are elicitor-induced signals transduced into cellular messages and regulated? The answers to these questions will help in the understanding of

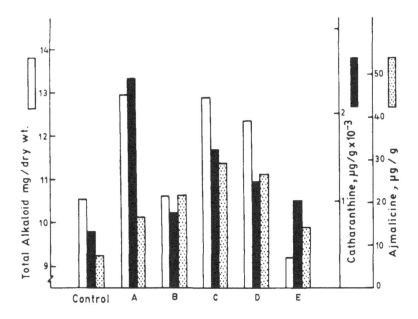

FIGURE 3.   Effect of bioregulators on alkaloid biosynthesis by cells of *Catharanthus roseus*.
(A) 1,1-dimethylpiperidine. (B) 2-diethylaminoethyl 3,4-dichlorophenylether. (C) 2-diethyl-
aminoethyl-2,4-dichlorophenylether. (D) 2-diethylaminoethyl-β-napthylether. (E)
2-diethylaminoethyl-3,4-dimethylphenylether. (From Lee, S. L., Chang, K. D., and Scott,
A. I., *Phytochemistry*, 20, 1841, 1981. With permission.)

induction and regulation of secondary biosynthesis. There is some evidence which shows
that the mechanism of elicitation is at the level of gene expression. Elicitation of phenylalanine acetate-derived secondary metabolites (isoflavonoids) is mediated by *de novo* mRNA
transcription and enzyme synthesis in cell culture of *Phaseolus vulgaris*[26,84] and *Petrose-
lenium hortense*.[79]

C. Biotransformation

Undifferentiated cell cultures of *Digitalis lantana* do not produce cardiac glycosides, but
they are able to perform special biotransformation on substrates added to the medium.[125]
Plants of *D. lantana* produce cardiac glycosides, digitoxin and digoxin, widely used for
treatment of heart ailments. Presently, digitoxin is relatively less in use than digoxin. This
has led to increasing stockpiles of digitoxin. Digitoxin is a byproduct in the commercial
production of digoxin from leaves of *D. lantana*. Digoxin differs from digitoxin only by an
additional hydroxyl function at C-12. Digitoxin can be methylated by chemical means to β-
methyldigitoxin which could be transformed by selected cell lines of *D. lantana* very rapidly
and efficiently to β-methyldigoxin (Figure 4), a widely used pharmaceutical.

Another instance of biotransformation is by cell lines of *Catharanthus roseus*. Vinblastine
and vincristine, the bis-indole alkaloids of this plant are employed to fight leukemia. Chemically synthesized 3,4-dehydrovinblastine was metabolized to leurosine, catharine, and several other bis-indole alkaloids by cell cultures of *C. roseus*.[81]

D. Cell Immobilization

It has been widely recorded that compact, organized, and slow-growing tissues accumulate
more alkaloids than the friable and rapidly growing cells. Therefore, it can be visualized
that the closer a cell or group of cells are to the whole plant, the more likely they are to be
similar in their metabolic pattern to that of whole plant. Hence, there is a need for cell

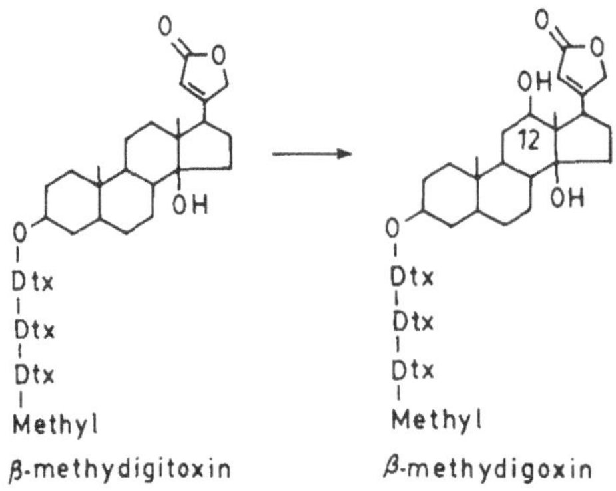

FIGURE 4. Biotransformation of β-methyldigitoxin to β-methyldigoxin by cells of *Digitalis lanatana.* (After Alfermann, A. W., Schuller, I., and Reinhard, E., *Planta Med.*, 40, 218, 1980. With permission.)

immobilization. In this technique, isolated cells are made to grow in such a way that it is a simulation of physical atmosphere found in a whole plant.

Immobilized cells are considered to be in an environment which is more conducive for production of desired metabolites.

1. Methods of Cell Immobilization

The first instance of cell immobilization was in lichen when cells of *Umbilicaria* were entrapped in polyacrylamide gel.[112] This was followed by growing of animal embryonic cells on DEAE-Sephadex microbeads.[161,162] Since then, a variety of cells/organisms have been made to grow in an immobilized state. The methods employed basically fall in four categories.

1. Immobilization by entrapment in an inert substratum, made out of a single or a combination of gels such as agar, agraose, alginate, chitosan, gelatin, polyacrylamide, or collagen. For first ever immobilization of plant cells, calcium alginate was employed to embed cells of *Morinda citrifolia, Catharanthus roseus,* and *Digitalis lantana.*[17]
2. Adsorption of cells to an inert substratum. In this method, more commonly employed for animal systems, cells are made to adhere to a charged surface which can be either microspheres or glass beads.
3. Adsorption to an inert substratum via a biological macromolecule such as lectin.
4. Covalent coupling of cells to an inert substratum such as carboxymethyl cellulose.

The last two methods are uncommon. The second one is more common for animal cells and first one alone is commonly employed for plant cells. The major advantages of the entrapment procedure are high cell density, mild immobilization conditions, and low risk of loss of cells from matrix.

For immobilization of plant cells, either agar or calcium alginate can be employed. The cells can be entrapped in small pieces of nylon netting made out of nylon pan scrubbers. When agar is to be employed, cells are at first mixed in molten agar at 35 to 40°C and sterile pieces of nylon netting are dipped in the suspension. When calcium alginate is to be employed, 2% solution is autoclaved and at room temperature cells and pieces of nylon netting are dipped in this solution. These pieces are then quickly transferred to 0.05 M sterile solution of $CaCl_2$ to allow calcium alginate to form a gel. The growth of cells entrapped in calcium alginate is better than in agar.

2. *Physiology of Immobilized Cells*

Cell immobilization per se allows a closer physical and chemical contact between cells that is not possible in an ordinary cell culture. Compared to a cell culture, growth of immobilized cells is much slower. The reduced growth rate of immobilized cells is the result of diffusion limitations and not the result of cell damage due to entrapment. Cells immobilized by the entrapment method are viable and biosynthetically active. The advantageous effect of slow growth, which is nearer to natural growth, can be seen in *Nicotiana* cells. An artificial inhibition of growth of callus cultures resulted in peroxidase pattern which was similar to that found at the onset of differentiation.[91] Slow growth leads to cell aggregation and it is one of the aspects of organization. The beneficial role of organization in synthesis of secondary metabolites has already been highlighted. Another way in which slow growth can be beneficial for synthesis of secondary metabolites is by way of competition for precursors between primary and secondary pathways. Theoretically, one can assume that if growth is slowed down, primary pathways can be blocked and secondary pathways will become operative and this might result in synthesis of secondary metabolites. Alternatively, if the lifespan of a mature nondividing cell culture is extended, it offers a means of increasing the yield of secondary products by cultured cells.[113] This was seen when the stationary phase culture of rose cells was treated with sucrose and spermidine, a treatment previously shown to arrest senescence; it resulted in accumulation of a wider range of phenolics than present in actively growing cells.

3. *Achievements of Plant Cell Immobilization*

Interest in immobilization of plant cells for synthesis of secondary metabolites such as alkaloids has been of recent origin.[17,19] Examples of achievements include *de novo* synthesis of anthraquinones (Figure 5) in *Morinda citrifolia,* which was increased ten times and prolonged by immobilization, and also a significant increase in production and excretion of ajamalicine by immobilized *Catharanthus roseus* cells.[17,18] However, stronger evidence is capsaicin production by immobilized cells of *Capsicum frutescens* which could be increased from nanogram quantities to hundreds or, exceptionally, thousands of micrograms by supplying 5 mM isocapric acid, a precursor of capsaicin.[180]

Cells immobilized in alginate were more efficient at capsaicin synthesis than cells in agar.[90] To an extent, the increased yield of capsaicin observed in cultures of entrapped pepper cells is associated with reduction in rate of protein synthesis as compared with suspended cells.[89,90]

Cells of *Digitalis lantana*[3] when immobilized in alginate gel biotransformed β-methyldigitoxin to β-methylodigoxin. It was possible in a small batch culture system. After a short lag period the immobilized cells could hydroxylate β-methyldigitoxin at a constant rate for more than 170 days.

Along with these encouraging reports on the role of cell immobilization, doubts have been expressed about the role of process. In addition to labor-intensive practice, the immobilized plant cell system is visualized to be comparable to a callus culture. Evidence for this is the unpublished work on cosmetic production from callus culture of *Jasmine officinale*.[27]

V. ORGANELLE DIFFERENTIATION — GREENING AND CELL PHOTOAUTOTROPHY

Higher plant cells in culture normally lose chlorophyll and are heterotrophic. However, it has been possible to establish actively growing cells and tissues with functional chloroplasts. Photoautotrophic cells of higher plants provide a new system for research on photosynthesis, since they are capable of utilizing solar energy. They are also reckoned as a potential source of increased productivity. As for differentiation, these cell cultures provide a suitable system

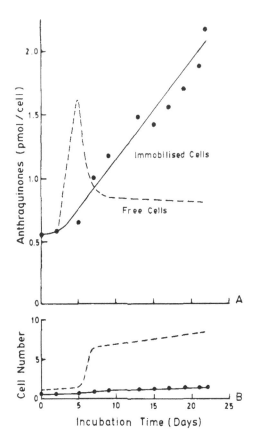

FIGURE 5.   Effect of immobilization on anthraquinone biosynthesis (A) and growth (B) of *Morinda citrifolia* cells. (From Brodelius, P., Deus, B., Mosbach, K., and Zenk, M. H., *FEBS Lett.*, 103, 93, 1979. With permission.)

for the study of structural and physiological events concerning differentiation of an organelle, i.e., transformation of a proplastid into chloroplast.

This research opens the new possibility of obtaining plants which can turn out to be more photosynthetically efficient than the original cultivars.

A. Index of Photoautotrophic Growth

The amount of chlorophyll present in a tissue has been taken to be the criterion of photoautotrophic growth. However, this is questionable because chlorophyll synthesis in the tissue of *Kalanchoe fedtschenkoi* was saturated at light intensity of 5000 lux, but for optimal photosynthetic functions (O_2 evolution, CO_2 fixation, and photosystem activities) higher light intensity of 12,000 to 14,000 lux was required.[136] Therefore, chlorophyll is not a suitable index of photosynthetic potential. Instead, in green tissue of *Hyoscyamus* a correlation has been found between photosynthetic activity and photoautotrophic growth (Figure 6) and such a correlation does not exist between photoautotrophic growth and chlorophyll content.[178]

B. Factors Favoring Chlorophyll Formation

Two major factors favoring chlorophyll synthesis in a tissue on culture are nutrients and cultural conditions.

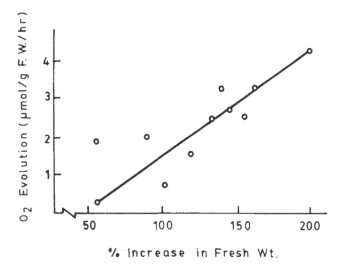

FIGURE 6. Relationship between photosynthetic activity and photoautotrophic growth of *Hyoscyamus* cells. (From Yamada, Y. and Sato, F., in *Handbook of Plant Cell Culture*, Evans, D. A., Sharp, W. R., Ammirato, P. V., and Yamada, Y., Eds., Macmillan, New York, 1983, 489. With permission.)

The nutrient conditions that affect chlorophyll synthesis are auxin, cytokinin, and sugar. Auxin, 2,4-D, commonly employed for initiation and multiplication of tissues is inhibitory to chlorophyll synthesis.[144] Therefore, NAA is preferred.[176] Cytokinin promotes greening in nongreen callus,[70] and at times there is synergism between cytokinin and sugar. In earlier literature sucrose has been described to be promotory for chlorophyll synthesis, but recently it has been shown that it inhibits photosynthetic activity and greening of tissue.[178] In photosynthetic cells of potato, sucrose inhibited photosynthetic activity and this could be separated from chlorophyll synthesis.[82]

Loss of chlorophyll in a tissue on continuous culture has been attributed to three factors:

1. Presence of sucrose in the medium.[35]
2. Growth in low light intensity or dark.[177]
3. Unbalanced growth[82] — cell division exceeds chlorophyll synthesis.

Cultural conditions affecting chlorophyll synthesis are light intensity[9] and gas phase.[25,30]

An adequate intensity of light (6000 to 12,000 lux) is required for transformation of amyloplasts into chloroplasts and photosynthetic functions.

Enhanced CO_2 concentration (1 to 2%) is an absolute requirement for photoautotrophic growth. High concentration of CO_2 is maintained by bubbling CO_2 enriched air in liquid culture. Alternately, petri dish culture or culture in a flask with cotton plug is maintained in a cabinet enriched with CO_2. However, it is not clear why a high CO_2 level is required. It is very likely that inabliity of cultured cells to grow photoautotrophically under ordinary air (0.1% CO_2) may be due to high CO_2 compensation point T, an index of photorespiration. T is higher in high CO_2-requiring cells than low CO_2-requiring cells.[158] It is higher for autotrophic and mixotrophic cells of tobacco than cells isolated from tobacco leaves. Also, dark respiration was higher than cells isolated from leaves. The high CO_2 compensation point of green cells in culture reflects a low concentration of CO_2 which may be due to low carbonic anhydrase activity or high activity of dark respiration. Therefore, instead of CO_2-enriched air, stimulation of carbonic anhydrase activity or suppression of dark respiration might help in the greening of tissue.

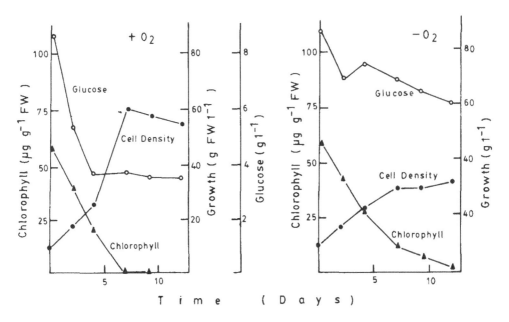

FIGURE 7.   Photoautotrophy of spinach cells in presence and absence of oxygen. (From Laulhere, J. P., Aguettaz, P., and Lescure, A. M., *Physiol. Veg.*, 22, 765, 1984. With permission.)

While describing the conditions for autotrophic growth of cells, the necessity of CO_2 has been emphasized, but there was no demonstration that growth of illuminated colonies is strictly dependent on CO_2 assimilation. This was, however, possible when instead of agar or agarose an inert solid support was provided to cells. When green cells of *N. tabacum* cv. Havana were incubated in CO_2-free air on a medium lacking sucrose, but solidified with 1% agar (three brands — Difco, Fischer® and Noble®) or agarose (two brands — Bio-red or Sea-Prep) an increase in relative dry weight was sustained through two passages. Instead, when polyurethane pads were employed to support the cells in liquid medium lacking sucrose, growth was negligible in two passages in CO_2-free air and it increased with each increment in CO_2 concentration.[96] Thus, it was possible to have a strict autotrophic culture of tobacco cells.

Low O_2 supply has also been described to be essential for photoautotrophic cells. However, it is not clear how it is beneficial. The level of O_2 in air can be lowered by adding N_2 gas.[28] In a recent investigation it was possible to obtain spinach cell suspension in active growth (Figure 7) with functional chloroplasts in the presence of glucose and in the absence of O_2.[83] It is very likely that when the exogenous oxygen supply is completely suppressed, respiration of glucose becomes dependent on availability of photosynthetic oxygen. In these conditions chloroplasts develop in the presence of glucose and cells divide actively. Thus, a fast-growing photomixotrophic cell suspension could be obtained.

C. Developmental Aspects of Greening

Structural changes in the differentiation of plastid during greening, a transition from proplastid to chloroplast, have been worked out in tobacco tissue.[15] On continuous culture in dark the cells became achlorophyllous. These cells, however, had amyloplasts which were devoid of lamellae. An exposure to light resulted in enrichment of cytoplasm accompanied by transformation of amyloplasts to proplastids. When these cells prepared to divide, dividing proplastids with rudimentary single lamellae were seen in them, typical of those seen in a meristematic cell from a plant. With the progress of cell division there was progressive differentiation of grana and intergrana thylakoids. By the time cells reached stationary phase the lamellar system was typical of a higher plant chloroplast.

D. Physiological and Biochemical Aspects of Greening

On transfer of tobacco tissue[116] from dark to light, striking features were an enormous increase in respiratory activity and a fourfold increase in soluble protein content. During multiplication of cells, an inverse relationship was noticed between phosphoenylpyruvate carboxylase (PEPCase) and ribulose bisphosphate carboxylase (RuBPCase). The former increased to twofold, and then returned to the original level in postexponential or stationary phase cells. The latter was minimal during the early stages of division and then reached a maximum, about a 20-fold increase. This increase was assigned to *de novo* synthesis. There was preferential synthesis of a large subunit of enzyme. Increased RuBPCase is also seen in green cells of *Digitalis purpurea*.[58,64]

A greening callus represents biochemical differentiation at the subcellular level. Therefore, changes related to organelle differentiation can be examined in this system. Cell suspension of *Nicotiana tabacum* cv. Wisconson 38 is a system in which chloroplast differentiation can be selectively induced (or repressed) under continuous illumination by addition (or omission) of cytokinin.[140] Green plastids from cytokinin medium and yellow plastids from cytokininless medium were able to support protein synthesis in vitro and differences were observed between the two peptide patterns.[87] These results indicate that cytokinin-dependent chloroplast development is accompanied by differential expression of plastome. Of late it is shown that there is a change in plastid mRNA population during cytokinin-induced chloroplast differentiation.[139] Also, there is evidence that plastid and nuclear DNA synthesis are not coupled.[60]

The photoreceptor system involved in chloroplast development responds to both blue and red light. However, in some cases only blue light induces and maintains the formation of chloroplasts while red light is ineffective. One such system is *N. tabacum* cv. Samsun where greening is strictly under blue light control.[10,128] The action of blue light is primarliy through plastid mRNA induction[127] (Figure 8).

There are also basic differences between nongreen and green cells in terms of energy-requiring and energy-generating processes. Soybean cells in suspension culture on continuous illumination and in the presence of glucose showed chloroplast biogenesis which was absent in cells maintained in dark. The ADP/ATP ratio and energy charge in both kinds of cells were equal from day 2 to 5 of culture. On day 6, the ADP/ATP ratio increased and the energy charge decreased while in white cells both parameters remained relatively constant. The failure of green cells to maintain equilbrium between energy-requiring and energy-generating processes might be related to energy demands for continued development of internal membrane system of chloroplasts.[75]

Chloroplasts are important in nitrogen metabolism. A follow up of nitrogen-assimilating enzymes during photoautotrophic growth of *Chenopodium rubrum* cells in suspension culture (which had a 3-day lag, 3-week log phase, and 10-day stationary phase) indicated that to begin with, ammonium was taken in preference to nitrate. However, during the second week of growth ammonia and nitrate were taken up simultaneously. Glutamine synthetase had a high specific activity in day 1 cells and this was sustained until mid-lag phase when it increased by 20%. Glutamate synthase activity increased rapidly in lag phase cells, but decreased at day 9 to about 50% and remained constant. Nitrate reductase activity increased rapidly in lag phase and reached a plateau that lasted from day 4 to 14. Nitrite reductase activity was high at day 5 and increased to a maximum on day 15. NADPH- and NADH-dependent glutamate dehydrogenase specific activities remained rather constant throughout the growth cycle.[21]

Activity of the enzyme glutamate synthase (GOGAT) is NADH and ferredoxin dependent. The physiological meaning of these two activities is not clearly understood. This could be resolved by taking nonchlorophyllous and chlorophyllous tissue of *Bouvardia ternifolia*.[114] Ferredoxin-GOGAT was found only in green tissues and was confined to chloroplast fraction. The activity increased parallel to the greening of tissue while NADH-GOGAT was not affected by the greening process.

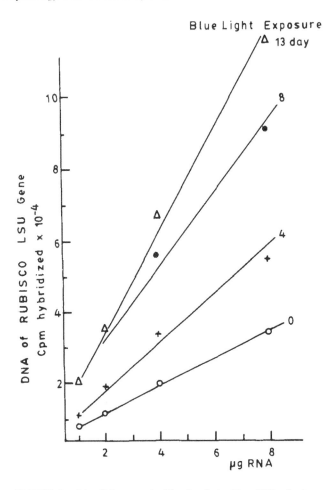

FIGURE 8.   Blue light control of levels of plastid m-RNA of tobacco cells in culture. (From Richter, G., *Plant Mol. Biol.*, 3, 271, 1984. With permission.)

E. Selection of Photoautotrophic Cells from Mixotrophic Cells

In the process of greening, not all cells become chlorophyllous and are fully autotrophic. Therefore, a selection of cell lines is made from mixotrophic cells. To begin with, the mixotrophic culture is grown on a sucrose medium and when the sugar is depleted, only those cells capable of growing photoautotrophically on a sugar-free medium survive. In this way photoautotrophic cell lines have been isolated from cells of *Chamaecereus sylvestrii*,[135] *Chenopodium rubrun*,[62] *Cytisus scoparius, Hyoscyamus niger, Nicotiana tabacum*,[158] *Asparagus*,[120] *Arachis hypogea*,[137] *Gisekia pharnacloides*,[138] and *Glycine max*.[61]

F. Induction of Photoautotrophic Cells

Instead of selection from mixotrophic cultures, a method is devised for induction of photoautotrophic cells.[179] In this method leaf segments are cultured on sugar-free medium containing salts, NAA (5 to 10 μM) and a cytokinin (BAP, 0.05 to 0.5 μM). The culture is continuously illuminated (3000 to 5000 lux) and maintained in a 20-ℓ jar aerated with 1% CO_2 at a flow rate of 1 ℓ/min. In such cultures, after 2 weeks small green calli are noticeable. These are photoautotrophic cells and remain so on separation. This method worked well for *Datura* and *Hyoscyamus*, but was ineffective for *Atropa*.

G. Continuous Culture of Photoautotrophic Cells

Sustained photoautotrophic growth in the form of callus culture has been possible in *Chenopodium rubrum* cells for over 2.5 years on regular subculture at 14-day intervals.[62,64,65] The same has been true of *Asparagus officinalis* cells for over 5 years.[120]

Continuous suspension cultures of photoautotrophic cells have also been possible in *Spinacea oleracea*,[28] *Asparagus*,[120] *Ocimum*,[29] and *Chenopodium*.[63]

VI. MORPHOLOGICAL DIFFERENTIATION — CYTODIFFERENTIATION

Differentiation of vascular elements, particularly tracheary elements, is described as cytodifferentiation. Cytodifferentiation occurs in apices, wound callus, callus tissue, cell suspension, and even in free cells. This topic has been a subject of several reviews[6,56,118,153,154,156] and even a book entitled *Cytodifferentiation in Plants — Xylogenesis as a Model System*[130] has appeared.

Various systems have been employed for the study of cytodifferentiation. Those falling within the scope of this text are callus tissue, free cell, protoplast, and intact tissues: pith cylinder and hypocotyl segment. In particular, parenchymatous cells are a better system. A cell or a tissue on culture can be subjected to a precise control of exogenous factor and excludes the correlations present in an intact plant. However, cells on culture do not depict the diversity of cytodifferentiation normally encountered in an intact plant. One of the reasons for delimiting different modes of cytodifferentiation is our inability to identify phloem elements and only tracheary elements can be easily detected by stains and stand macertation for quantification. Therefore, for all practical purposes and here also, cytodifferentiation refers to differentiation of tracheary elements.

A. Cytodifferentiation in Callus Tissue

A basis for cytodifferentiation was revealed when a graft of a vegetative bud on the upper surface of a callus tissue of *Cichorium* resulted in differentiation of vascular elements below the region of bud in the callus.[22] This also happened even when the physical contact between the callus and bud was severed by placing cellophane paper at the site of graft, indicating that the stimulus for cytodifferentiation is diffusable and chemical in nature. This result was confirmed when tissue of *Syringa vulgaris*[168] was employed as an experimental system. By making a V-shaped incision on the callus, a vegetative bud was placed in the cavity. To make a proper union between bud and callus cells, nonnutrient agar was placed in the cavity. Within 20 to 30 days differentiation of vascular elements could be seen in the callus below the region of the bud (Figure 9). A more revealing aspect of this experiment was that agar containing sucrose and auxin could effectively substitute the stimulus from the bud for the induction of vascular tissue differentiation.

Control of cytodifferentiation was possible in callus tissues of *Syringa vulgaris*[167] and *Phaseolus vulgaris*[66] in the sense that there was no correlation between auxin and sucrose, and sucrose alone had a predominant role. At an auxin level of 0.05 mg/ℓ and 1% sucrose, only a few xylem elements could be seen in *Syringa* callus. However, 1.5 to 2.5% sucrose induced strong xylem differentiation with little or no phloem as long as a small quantity of auxin (0.05 mg/ℓ) IAA was available. By contrast, differentiation of phloem with little or no xylem was possible at 3 to 4% sucrose. Intermediate level of sucrose 2.5 to 3.5% favored formation of xylem as well as phloem. The regulatory role of sucrose was also described for *Phaseolus*[66] tissue. Of the sugars, monosaccharides were nonstimulatory, but sucrose, maltose, and trehalose were effective. This led to the thinking that sucrose has some hormonal influence.

Further work on control of cytodifferentiation is, however, contradictory to the findings reported above. Higher levels of sucrose increased xylem differentiation in callus of

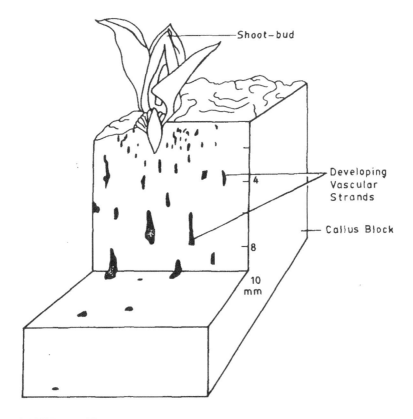

FIGURE 9.   Differentiation of vascular elements in callus tissue of syringa on grafting a shoot-bud bearing two to four leaf primordia. (After Wetmore, R. H. and Sorokin, S., *J. Arnold Arbor.*, 36, 305, 1955. With permission.)

Parthenocissus[129] in the presence of auxin. In a reinvestigation[4] of work on *Syringa* along with other genera *Daucus* and *Glycine*, low levels of IAA resulted in differentiation of sieve elements with no tracheary cells. High levels resulted in both phloem and xylem. Thus, IAA controlled the number of sieve and tracheary elements. Changes in sucrose concentration, keeping IAA level constant, did not have specific effect either on sieve element differentiation or on ratio between phloem and xylem. Increase in sucrose, however, increased the deposition of callose on sieve plates. Therefore, the interdependence of auxin and sugar remains to be resolved.

Other hormones that are implied in the control of cytodifferentiation are cytokinin and gibberellin. In tobacco[8] tissue, addition of kinetin $10^{-7}M$ to medium containing 5% coconut milk enhanced tracheidal differentiation by 30%. Similarly, in *Glycine*[43] tissue, cytokinin stimulated cytodifferentiation. A synergism between IAA and kinetin is seen in tobacco tissue.[8] Contrary to these reports, cytokinin did not stimulate xylogenesis in *Helianthus*[109] and *Linum*.[98] The role of cytokinin in xylogenesis, however, became apparent in conjunction with phosphate. Medium with high concentration of benzylaminopurine (BAP) and low concentration of phosphate induced the formation of tracheary elements in suspension cultures of *Asparagus plumosis*,[2] whereas a low level of BAP and low levels of phosphate induced the formation of parenchyma cells.

As for gibberellin, there are reports both for and against its stimulation of cytodifferentiation. Gibberellins, either autoclaved or filter sterilized are inhibitory to xylogenesis in *Helianthus tuberosus*.[165]

A new note in control of cytodifferentiation was introduced when the role of ethylene became evident[105] in tissues of soybean. The seedlings of soybean cultivar Clark 63 show abnormal growth at 25°C due to excessive ethylene biosynthesis: at this temperature, cultivar Wayne is normal. The callus tissue from Wayne showed increased xylogenesis in response to exogenous methionine, in comparison to IAA-K controls, at both 20 and 25°C, whereas tissue from Clark 63 showed increased xylogenesis in response to exogenous methionine only at 25°C. Further, xylogenesis initiated by IAA-K combine was inhibited by silvernitrate (20 μg/ℓ) to the extent of 75% in Wayne tissue and only 6% in the tissue from Clark 63. The inhibition by AgNO₃ was partially reversed by addition of methionine. Also, in lettuce pith explants ethylene is involved in IAA-kinetin-induced initiation of xylem differentiation.[106] Basically, ethylene is a senescence-promoting substance for plant cells.[124,175]

The key enzymes of lignin biosynthesis are phenylalanine ammonia lyase (PAL) and peroxidases (PO). Xylem differentiation in cultured tissues has been correlated with increased levels of PAL[51,57,133] and wall-bound PO.[51] Another work points towards the appearance of new acid phosphatase during cytodifferentiation in *Vigna*.[33] The function of ethylene in control of lignification during xylogenesis is ascribed to induction of activity of wall bound PO[104] because addition of silver, an ethylene antagonist, to the culture medium inhibited lignification and markedly reduced wall-bound PO activity. Exogenous L-methionine, an ethylene precursor, completely reversed the inhibitory effect of silver on lignification and wall-bound PO activity.

Evidence has also been presented for proteins specific for differentiation of vascular elements in maize.[74] In 13- to 15-day-old developing embryos appear specific stellar antigens which were also found in embryo- or stem-derived cell suspension.

B. Is Cell Division a Prerequisite for Cytodifferentiation?

In most developmental functions a period of cell division precedes differentiation and treatments that inhibit cell division also inhibit differentiation. Therefore, it is quite in order to inquire whether or not cell division is a prerequisite for cytodifferentiation, particularly when factors regulating cell division and cytodifferentiation are the same.

The first evidence for cell division to be a prerequisite for cytodifferentiation was provided when fluorodeoxyuridine (FUDR) or mitomycin-C (MC) inhibited xylogenesis in stem segments of *Coleus*.[42] This was followed by a work on pea root segments which showed that FUDR prevented cytodifferentiation and cell division and BUDR (bromodeoxyuridine) specificially prevented cytodifferentiation with little effect on cell division.[141] Labeling studies of tuber tissue of *Helianthus* employing tritiated thymidine also provided the evidence that cell division precedes cytodifferentiation; characteristically sculptured cell wall appears after 6 to 8 hr of the completion of last division.[92]

Contrary to the above findings, an indication that cytodifferentiation is possible without cell division was provided in experiments on lettuce explants which formed a few tracheidal elements (2.6% of control) in the presence of FUDR.[159] Also, in tuber tissue of *Helianthus*[122] direct differentiation was possible if the explants were taken from immature developing tubers. When division of cells in these explants was inhibited by prior exposure of tubers to 20 krad of gamma radiation, cytodifferentiation proceeded on an auxin-containing medium. The capacity for direct differentiation, however, declined with tuber maturity. Therefore, relative maturity of a cell is an important factor determining direct differentiation. Inhibition of differentiation of tracheary elements without affecting cell division is possible by 3-aminobenzamide,[59] a specific inhibitor of nuclear enzyme ADP-ribosyl transferase. This inhibitor blocked cytodifferentiation (Figure 10) in cells of Jerusalem artichoke during a 6-hr period, prior to the onset of visible differentiation.[123] The blocking of cytodifferentiation by the inhibitor was reversed on removal.

Despite these evidences, a muticellular tissue is not a suitable system to settle this question.

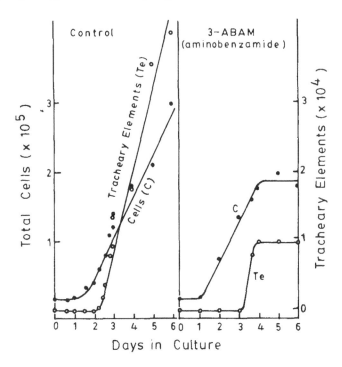

FIGURE 10.   Inhibition of tracheary element differentiation by 3-aminobenzamide. (From Philips, R. and Hawkins, S. W., *J. Exp. Bot.*, 36, 119, 1985. With permission.)

C. Cytodifferentiation in Free Cells

The observations that differentiation of isolated cells is possible into tracheidal elements lend strong support to the idea that cell division is not a prerequisite for cytodifferentiation.

Cells of *Centaurea*[155] in agar plates differentiated directly into tracheids without a preceding cell division. These cultures were, however, raised from actively growing cell suspension; thus, an element of doubt remained that these cells were the products of recent division. More convincing evidence was provided when isolated mesophyll cells of *Zinnia elegans* differentiated into tracheary elements after a period of extension, but without a division.[76] Also, regular observations of single cells of *Zinnia* were recorded photographically and they were shown to undergo cytodifferentiation without preceding cell division.[48] Colchicine, which inhibited cell division in the system, did not impair differentiation.

Isolated mesophyll cells of *Z. elegans* can be induced to differentiate into tracheids at a frequency of up to 30% of the cell population, after 96 hr of culture, with over 50% of these cells developing into tracheids without a preceding cell division.[48] For direct differentiation, cells from young developing leaves responded more rapidly and in high frequency than cells from older leaves (Figure 11). Also the protoplasts of these cells differentiated directly into tracheids.[77] These features make *Zinnia* a model system for studies on cytodifferentiation.

It is generally believed that changes in the pattern of gene expression that lead to differentiation occur at some stage of the cell cycle. Therefore, it is pertinent to know — is cell cycle activity necessary for cytodifferentiation?[34] Surprisingly, cells of *Zinnia* differentiated into tracheids without intervening DNA replication.[49] More than 55% of the nuclei of immature tracheary elements were at 2 C level of DNA and were not labeled by continuous feeding with tritiated thymidine. However, inhibitors of DNA synthesis prevented cytodifferentiation. It is via blockage of synthesis of some DNA, since replication of whole genome is not essential for this type of differentiation.[50]

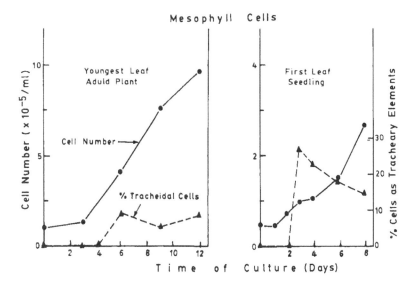

FIGURE 11.   Frequency of differentiation of tracheidal cells from cells of *Zinnia elegans*. (After Fukuda, H. and Komamine, A., *Plant Physiol.*, 65, 61, 1980. With permission.)

If not DNA synthesis, is transcription and translation required for cytodifferentiation? The answer to this question is in the affirmative[52] (Figure 12). Actinomycin-D prevented cytodifferentiation in isolated cells of *Zinnia* and administration of the drug from 24 to 60 hr of culture was most effective; at this time, incorporation of ^{14}C uridine into nucleic acid was at its peak. Cycloheximide also prevented cytodifferentiation; this was more pronounced when the drug was given between 24 and 60 hr of culture. Incorporation of ^{14}C leucine into protein was also higher during this period. These results clearly indicate that RNA and protein synthesis are prerequisites for cytodifferentiation. Further, it was resolved that synthesis of two specific polypeptides was shut off and two new polypeptides were synthesized between 46 to 60 hr preceding the morphological changes leading to cytodifferentiation. An analysis of PO isozymes, in particular, revealed that a new cathodic PO appeared in cells during cytodifferentiation.[94] The process of cytodifferentiation is inhibited by triiodobenzoic acid (TIBA), an inhibitor of auxin.[20]

1. Developmental Aspects

The process of tracheary element formation by individual mesophyll cells of *Zinnia elegans* has been followed employing scanning and transmission electron microscopy.[20] A freshly isolated cell has a thin primary wall, starch-filled chloroplasts, and a nucleus with a small nucleolus and peripherally located chromatin. During culture, the nucleolus expands and becomes vacuolate and the chromatin in the nucleus becomes dispersed. Wall thickening is accompanied by an increase in dictyosome activity and localization of microtubules over the thickenings. Wall hydrolysis results in granular reticulate appearance.

Relatively little is known about the precise involvement of cytoskeletal elements in tracheary element differentiation. Employing *Zinnia* mesophyll cells, a monitoring of cytoskeletal elements was done before and during the appearance of secondary walls. The first microscopically visible sign of secondary wall formation involved the reorganization of microtubules from cortical arrays into distinct groups.[40] In particular, cortical microtubules reorient from longitudinal to transverse arrays as the differentiation of tracheary elements becomes visible. The orientation of microtubules determines the direction of secondary wall bands. Application of taxol to culture stabilized the microtubules of most cells in the longitudinal direction.[41] Tracheary element differentiation in taxol-treated cultures showed sec-

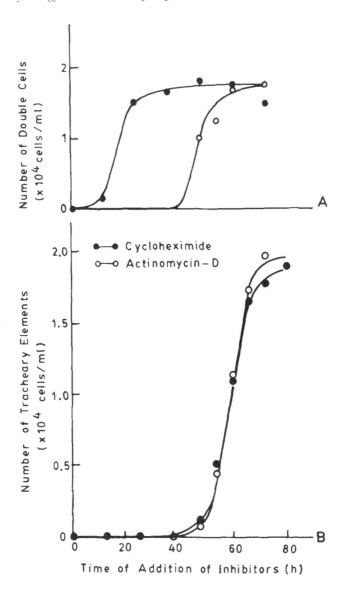

FIGURE 12.   Effect of inhibitors on differentiation of tracheidal elements from mesophyll cells of *Zinnia elegans*. (After Fukuda, H. and Komamine, A., *Plant Cell Physiol.*, 24, 603, 1983. With permission.)

ondary wall bands parallel to the long axis of cell while in control cultures, the wall bands were transverse to the long axis of cell.

D. Cytodifferentiation in an Intact Tissue

From studies on cytodifferentiation in a callus tissue it can be concluded that the need of auxin is well established. Sucrose is also required, but an interdependence of sucrose and auxin needs to be resolved. Cytokinin is necessary for xylogenesis in some, but not all tissues. In a multicellular and amorphous system, like a callus, it is difficult to pinpoint the role of each morphogenic substance in control of cytodifferentiation,[39] particularly when callus tissue shows dependence for auxin for its growth and is independent of cytokinin requirement. Therefore, of late, the problem has been approached by taking tissue blocks, either explants from pith or segments from hypocotyl.

By taking explants from pith of lettuce and applying the morphogenic substances at opposite ends of tissue it was possible to know the mobility of substance and its role in callus formation and tracheary element induction.[174] Callus formation was stimulated by IAA and was limited to a few millimeters at the ends of explant. Tracheary elements appeared only in callus and for induction both IAA and zeatin were required. IAA and kinetin were effective only close to the site of application whereas sucrose diffused the 10-mm length of explants. In absence of sucrose some tracheary elements appeared, but increased with the application of sucrose. This work reveals the necessity of auxin, cytokinin, and sucrose in cytodifferentiation.

For the differentiation of secondary xylem fibers in hypocotyl segments of *Helianthus annus*[5] to begin with cytokinin, gibberellin, and auxin are required, but later stages of fiber differentiation can occur in the absence of cytokinin.

The mode of action of cytokinin in cytodifferentiation is possibly via ethylene formation.[107] This is evidenced in work on pith explants of lettuce. Cytodifferentiation occurred when explants were supplied with IAA in combination with either an ethylene biosynthetic precursor ACC (1-amino-cyclopropane-1-carboxylic acid), or ethylene-releasing agent CEPA (2-chloroethyl phosphonic acid), or kinetin. However, no xylem elements appeared in the presence of individual substances like IAA, kinetin, ACC, and CEPA or when kinetin was given along with ACC or CEPA. Therefore, both ethylene and auxin are required for xylogenesis.

REFERENCES

1. **Ahuja, M. R.,** A cytological study of heritable tumors in *Nicotiana* species hybrids, *Genetics,* 47, 865, 1962.
2. **Albinger, G. and Beiderbeck, R.,** Differenzierung von Tracheiden und Speicterparenchymzellen in suspension kulturen von *Asparagus plumosus, Z. Pflanzenphysiol.,* 112, 443, 1983.
3. **Alfermann, A. W., Schuller, I., and Reinhard, E.,** Biotransformation of cardiac glycosides by immobilized cells of *Digitalis lantana, Planta Med.,* 40, 218, 1980.
4. **Aloni, R.,** Role of auxin and sucrose in the differentiation of sieve and tracheary elements in plant tissue cultures, *Planta,* 150, 255, 1980.
5. **Aloni, R.,** Role of cytokinin in differentiation of secondary xylem fibers, *Plant Physiol.,* 70, 1631, 1982.
6. **Barnett, J. R.,** *Xylem Cell Development,* Castle House Publications, London, 1981.
7. **Bayer, M. H. and Ahuja, M. R.,** Tumour formation in *Nicotiana,* auxin levels and auxin inhibitors in normal and tumour-prone genotypes, *Planta,* 72, 292, 1968.
8. **Bergmann, L.,** Der Einfluß von kinetin auf die Ligninbildung und differenzierung in Gewebkulturne von *Nicotiana tabacum, Planta,* 62, 221, 1964.
9. **Bergmann, L. and Balz, A.,** Der Einfluss von Farblicht auf wachstum und Zusammensetzung Pflanzlicher Gewebekulturen. I. *Nicotiana tabacum* var. "Samsun", *Planta,* 79, 285, 1966.
10. **Bergmann, L. and Berger, Ch.,** Farblicht und plastiden-differenzierung von zellkulturen von *Nicotiana tabacum* var. Samsun, *Planta,* 69, 58, 1966.
11. **Bezdek, M. and Vyskot, B.,** Spontaneous 5-bromodeoxyuridine tolerance of *Nicotiana* genetic tumors, *Biochem. Physiol. Pflanzen.,* 174, 479, 1979.
12. **Bezdek, M., Vyskot, B., Tkadlecek, L., and Karpfel, Z.,** Spontaneous high level Budr tolerance of unorganized crown gall tumour, *Biochem. Physiol. Pflanzen.,* 171, 537, 1977.
13. **Bostock, R. M., Laine, R. A., and Kuc, J. A.,** Factors affecting the elicitation of sesquiterpenoid phytoalexin accumulation by Eicosapentaenoic and arachidonic acids in potato, *Plant Physiol.,* 70, 1417, 1982.
14. **Boulanger, D., Bailey, B. K., and Steck, W.,** Formation of edulinine and furoquinoline alkaloids from quinoline derivatives by cell suspension cultures of *Ruta graveolens, Phytochemistry,* 12, 2399, 1973.
15. **Brangeon, J. and Nato, A.,** Heterotrophic tobacco cell cultures during greening. I. Chloroplast and cell development, *Physiol. Plant.,* 53, 327, 1981.
16. **Braun, A. C.,** Studies on the tumour inception in the crown gall disease, *Am. J. Bot.,* 30, 674, 1943.

17. **Brodelius, P., Deus, B., Mosbach, K., and Zenk, M. H.,** Immobilized plant cells for the production and transformation of natural products, *FEBS Lett.*, 103, 93, 1979.

18. **Brodelius, P. and Mosbach, K.,** *Chemical Tech. Biotechnol.*, 32, 330, 1982.

19. **Brodelius, P. and Nilsson, K.,** Entrapment of plant cells in different matrices, *FEBS Lett.*, 122, 312, 1980.

20. **Burgess, J. and Linstead, P.,** In vitro tracheary element formation: structural studies and the effect of triiodobenzoic acid, *Planta*, 160, 481, 1984.

21. **Campbell, W. H., Ziegler, P., and Beck, E.,** Development of nitrogen assimilation enzymes during photoautotrophic growth of *Chenopodium rubrum* suspension cultures, *Plant Physiol.*, 74, 947, 1984.

22. **Camus, C.,** Recherches sur le rôle des bourgeons dans les phenomenes de morphogenes, *Rev. Cytol. Biol. Veg.*, 9, 1, 1949.

23. **Capelle, S. C., Mok, D. W. S., Kischner, S. C., and Mok, M. C.,** Effects of thidiazuron on cytokinin autonomy and the metabolism of N^6-(Δ^2-isopentyl) (8C^{14}) adenosine in callus tissue of *Phaseolus lunatus*, *Plant Physiol.*, 73, 796, 1983.

24. **Chaturvedi, H. C. and Chaudhary, A. C.,** Habituation in root callus of *Citrus aurantifolia*. Free amino acid contents of normal and habituated callus, *Indian J. Exp. Biol.*, 15, 581, 1977.

25. **Corduan, G.,** Autotrophe Gewebekkulturen von *Ruta graveolens* und deren $^{14}CO_2$-Markierungs-Produkte, *Planta*, 91, 291, 1970.

26. **Cramer, C. L., Ryder, T. B., Bell, J. N., and Lamb, C. J.,** Rapid switching of plant gene expression induced by fungal elicitor, *Science*, 227, 1240, 1985.

27. **Dainty, A. L., Goulding, K. H., Robinson, P. K., Simpkins, I., and Trevan, M. D.,** Effect of immobilization on plant cell physiology — real or imaginary, *Trends Biotechnol.*, 3, 59, 1985.

28. **Dalton, C. C.,** Photoautotrophy of spinach cells in continuous culture: photosynthetic development and sustained photoautotrophic growth, *J. Exp. Bot.*, 31, 791, 1980.

29. **Dalton, C. C.,** Chlorophyll production in fed-batch cultures of *Ocimum basilieum* (sweet basil), *Plant Sci. Lett.*, 32, 263, 1983.

30. **Dalton, C. C. and Street, H. E.,** The role of gas phase in the greening and growth of illuminated cell suspension culture of spinach *(Spinacea oleracea)*, *In Vitro*, 12, 485, 1976.

31. **Darvill, A. G. and Albersheim, P.,** Phytoalexins and their elicitors — a defence against microbial infection in plants, *Annu. Rev. Plant Physiol.*, 35, 243, 1984.

32. **De Greef, D. and Jacobs, M.,** In vitro culture of the sugar beet: description of a cell line with high regenerative capacity, *Plant Sci. Lett.*, 17, 55, 1979.

33. **De, K. K. and Roy, S. C.,** Role of an acid phosphatase isoenzyme in callus tissue during cytodifferentiation, *Theor. Appl. Genet.*, 68, 285, 1984.

34. **Dodds, J. H.,** Is cell cycle activity necessary for xylem cell differentiation?, *What's New Plant Physiol.*, 10, 13, 1979.

35. **Edelman, J. and Hanson, A. D.,** Sucrose suppression of chlorophyll synthesis in carrot tissue cultures, *Planta*, 98, 150, 1971.

36. **Eilert, U., Ehmke, A., and Wolters, B.,** Elicitor-induced accumulation of acridine alkaloid epoxides in *Ruta graveolens* suspension culture, *Planta Med.*, 50, 508, 1984.

37. **Einset, J. W.,** Two effects of cytokinin on the auxin requirement of tobacco callus cultures, *Plant Physiol.*, 59, 45, 1977.

38. **Everett, N. P.,** 2,4-D independent cell cultures of sycamore *(Acer pseudoplatanus)*. Isolation, responses to 2,4-D and kinetin and sensitivities to antimetabolites, *J. Exp. Bot.*, 32, 171, 1981.

39. **Fadia, V. P. and Mehta, A. R.,** Tissue culture studies on cucurbits: the effects of NAA, sucrose, and kinetin on tracheal differentiation in *Cucumis* tissue cultured in vitro, *Phytomorphology*, 23, 212, 1973.

40. **Falconer, M. M. and Seagull, R. W.,** Immunofluorescent and calcofluor-white staining of developing tracheary elements in *Zinnia elegans* suspension cultures, *Protoplasma*, 125, 190, 1985.

41. **Falconer, M. M. and Seagull, R. W.,** Xylogenesis in tissue culture: taxol effects on microtubule reorientation and lateral association in differentiating cells, *Protoplasma*, 128, 157, 1985.

42. **Fosket, D. E.,** Cell division and the differentiation of wound vessel members in cultured stem segments of *Coleus*, *Proc. Natl. Acad. Sci. U.S.A.*, 59, 1089, 1968.

43. **Fosket, D. E. and Torrey, J. G.,** Hormonal control of cell proliferation and xylem differentiation in cultured tissues of *Glycine max* var. Biloxi, *Plant Physiol.*, 44, 871, 1969.

44. **Fowler, M. W.,** Commercial applications and economic aspects of mass plant cell culture, in *Plant Biotechnology*, Mantell, S. H. and Smith, H., Eds., Cambridge University Press, Cambridge, 1983, 3.

45. **Fox, J. E.,** Growth factor requirements and chromosome number in tobacco tissue cultures, *Physiol. Plant.*, 16, 793, 1963.

46. **Fox, E. J., Chen, C. M., and Gillam, I.,** Glucose catabolism in normal and autonomous tobacco tissue cultures, *Plant Physiol.*, 40, 529, 1965.

47. **Fujita, Y.,** Efficient production of shikonin derivatives by cell suspension cultures of *Lithospermum erythrorhizon*, *In Vitro*, 21 (Part II), 60A, 1985.

48. **Fukuda, H. and Komamine, A.**, Direct evidence for cytodifferentiation to tracheary elements without intervening mitosis in a culture of single cells isolated from mesophyll of *Zinnia elegans, Plant Physiol.,* 65, 61, 1980.

49. **Fukuda, H. and Komamine, A.**, Relationship between tracheary element differentiation and the cell cycle in single cells isolated from the mesophyll of *Zinnia elegans, Physiol. Plant.,* 52, 423, 1981.

50. **Fukuda, H. and Komamine, A.**, Relationship between tracheary element differentiation and DNA synthesis in single cells isolated from the mesophyll of *Zinnia elegans* — analysis by inhibitors of DNA synthesis, *Plant Cell Physiol.,* 22, 41, 1981.

51. **Fukuda, H. and Komamine, A.**, Lignin synthesis and its related enzymes as markers of tracheary element differentiation in single cells isolated from mesophyll of *Zinnia elegans, Planta,* 155, 423, 1982.

52. **Fukuda, H. and Komamine, A.**, Changes in the synthesis of RNA and protein during tracheary element differentiation in single cells isolated from the mesophyll of *Zinnia elegans, Plant Cell Physiol.,* 24, 603, 1983.

53. **Furuya, T. and Ishii, T.**, The Manufacturer of Panax Plant Tissue Culture Containing Crude Saponins and Crude Sapogenins Which are Identical with Those of Natural Panax Roots, Japanese Patent No. 48, 31917, 1972.

54. **Gamburg, K. Z.**, Factors affecting differentiation of plant tissues grown in vitro, in *Cell Differentiation and Morphogenesis,* Beerman, W., Ed., North Holland, Amsterdam, 1966, 53.

55. **Gamburg, K. Z. and Rekoslavyskaya, N. I.**, Effect of antiauxins on growth of isolated cultures of normal and autonomous plant tissues, *Dokl. Bot. Sci.,* 232, 7, 1977.

56. **Gautheret, R. J.**, Sur la culture de trois types de tissus de Scorsoniere tissus normaux, tissus de crown gall et tissus accountume a l'heteroauxin, *C. R. Acad. Sci.,* 226, 270, 1948.

57. **Haddon, L. and Northcote, D. H.**, Correlation of the induction of various enzymes, concerned with phenylpropanoid and lignin synthesis during differentiation of bean callus *Phaseolus vulgaris, Planta,* 128, 255, 1976.

58. **Hagimori, M., Matsumoto, T., and Mikami, Y.**, Photoautotrophic culture of undifferentiated cells and shoot-forming cultures of *Digitalis purpurea, Plant Cell Physiol.,* 25, 1009, 1984.

59. **Hawkins, S. W. and Phillips, R.**, 3-aminobenzamide inhibits tracheary element differentiation but not cell division in cultured explants of *Helianthus tuberosus, Plant Sci. Lett.,* 32, 221, 1983.

60. **Heinhorst, S., Cannon, G., and Weissbach, A.**, Plastid and nuclear DNA synthesis are not coupled in suspension cells of *Nicotiana tabacum, Plant Mol. Biol.,* 4, 3, 1985.

61. **Horn, M. E., Sherrard, J. H., and Widholm, J. M.**, Photoautotrophic growth of soybean cells in suspension cultures, *Plant Physiol.,* 72, 426, 1983.

62. **Husemann, W.**, Growth characteristics of hormones and vitamin independent photoautotrophic cell suspension cultures from *Chenopodium rubrum, Protoplasma,* 109, 415, 1981.

63. **Husemann, W.**, Continuous culture growth of photoautotrophic cell suspensions from *Chenopodium rubrum, Plant Cell Rep.,* 2, 59, 1983.

64. **Husemann, W., Herzbeck, H., and Robenek, H.**, Photosynthesis and carbon metabolism in photoautotrophic cell suspensions of *Chenopodium rubrum* from different phases of batch growth, *Physiol. Plant.,* 62, 349, 1984.

65. **Husemann, W., Plohr, A., and Barz, W.**, Photosynthetic characteristics of photomixotrophic and photoautotrophic cell suspension cultures of *Chenopodium rubrum, Protoplasma,* 100, 101, 1979.

66. **Jeffs, R. A. and Northcote, D. H.**, The influence of indole-3yl-acetic acid and agar on the pattern of induced differentiation in plant tissue culture, *J. Cell. Sci.,* 2, 77, 1967.

67. **Kamimura, S. and Nishikawa, M.**, Growth and alkaloid production of the cultured cells of *Papaver bracteatum, Agric. Biol. Chem.,* 40, 907, 1976.

68. **Kandra, G. and Maliga, P.**, Is bromodeoxyuridine resistance a consequence of cytokinin habituation in *Nicotiana tabacum, Planta,* 133, 131, 1979.

69. **Kaul, B. and Staba, E. J.**, *Dioscorea* tissue cultures. I. Biosynthesis and isolation of diosgenin from *Dioscorea deltoidea* callus and suspension cells, *Lloydia,* 31, 171, 1968.

70. **Kaul, K. and Sabharwal, P. S.**, Effects of sucrose and kinetin on growth and chlorophyll synthesis in tobacco tissue cultures, *Plant Physiol.,* 47, 691, 1971.

71. **Kehr, A. E. and Smith, H. H.** Genetic tumours in *Nicotiana* hybrids, in *Abnormal and Pathological Plant Growth,* Brookhaven National Laboratory, Upton, N.Y., 1954; *Brookhaven Symp. Quant. Biol.,* 6, 57, 1954.

72. **Kevers, C., Coumans, M., DeGreef, W., Hofinger, M., and Gaspar, Th.**, Habituation in sugarbeet callus: auxin content, auxin protectors, peroxidase pattern and inhibitors, *Physiol. Plant.,* 51, 281, 1981.

73. **Khanna, P. K. and Mohan, S.**, Isolation and identification of diosgenin and sterols from fruits and in vitro cultures of *Momordica charantia, Indian. J. Exp. Biol.,* 11, 58, 1973.

74. **Khavkin, E. E., Markov, E. Yu., and Misharin, S. I.**, Evidence for proteins specific for vascular elements in intact and cultured tissues and cells of maize, *Planta,* 148, 116, 1980.

75. **de Klerk-Kiebert, Y. M. and Van der Plas, L. H. W.,** The course of adenine nucleotide contents in white and green soybean cells during growth in batch suspension culture, *Plant Sci. Lett.,* 33, 155, 1984.

76. **Kohlenbach, H. W. and Schmidt, B.,** Cytodifferenzierung in Form einer direkten Unwandlung isolierter Mesophyll zellen zu Tracheiden, *Z. Pflanzenphysiol.,* 75, 369, 1975.

77. **Kohlenbach, H. W. and Schopke, C.,** Cytodifferentiation to tracheary elements from isolated mesophyll protoplasts of *Zinnia elegans, Naturwissenschaften,* 68, 576, 1981.

78. **Koves, E. and Sazbo, M.,** Some questions of auxin autotrophy in tobacco callus cultures, *Wiss. Z. Ernst-Moritz-Arndt-Univ. Greifsw. Math. Naturwiss. Reihe,* 19, 39, 1980.

79. **Kuhn, D. N., Chappal, J., Boudet, A., and Hahlbrock, K.,** Induction of phenylalanine ammonia-lyase and 4-coumarate: COA ligase in RNAs in cultured plant cells by UV light or fungal elicitor, *Proc. Natl. Acad. Sci. U.S.A.,* 81, 1102, 1984.

80. **Kurosaki, F. and Nishi, A.,** Isolation and antimicrobial activity of the phytoalexin 6-methoxymallein from cultured carrot cells, *Phytochemistry,* 22, 669, 1983.

81. **Kutney, J. P., Aweryn, B., Choi, L. N. G., Kolodziejezyk, N., Kurz, W. G. W., Chatson, K. B., and Constabel, F.,** Alkaloid production in *Catharanthus roseus* cell cultures. IX. Biotransformation studies with 3,4-dihydrovinblastine, *Heterocycles,* 16, 1169, 1981.

82. **LaRosa, P. C., Hasegawa, P. M., and Bressan, R. A.,** Photoautotrophic potato cells, transition from heterotrophic to autotrophic growth, *Physiol. Plant.,* 61, 279, 1984.

83. **Laulhere, J. P., Aguettaz, P., and Lescure, A. M.,** Regulation of the oxygen exchanges and of greening by controlled supplies of sugar in photomixotrophic spinach cell suspensions, *Physiol. Veg.,* 22, 765, 1984.

84. **Lawton, M. A., Dixon, R. A., Halbrock, K., and Lamb, C. J.,** Elicitor induction of mRNA activity: rapid effects of elicitor on phenylalanine ammonia-lyase and chalcone synthase mRNA activities in bean cells, *Eur. J. Biochem.,* 130, 131, 1983.

85. **LeGoff, L., Roussaux, J., Aaron-da Cunha, M. I., and Beljanski, M.,** Growth inhibition of crown-gall tissue in relation to the structure and activity of DNA, *Physiol. Plant.,* 64, 177, 1985.

86. **Lee, S. L., Cheng, K. D., and Scott, A. I.,** Effects of bioregulators on indole alkaloid biosynthesis in *Catharanthus roseus* cell culture, *Phytochemistry,* 20, 1841, 1981.

87. **Lescure, A. M.,** Chloroplast differentiation in cultured tobacco cells in vitro. Protein synthesis efficiency of plastids at various stages of their evolution, *Cell Differ.,* 7, 139, 1978.

88. **Lescure, A. M. and Peaud-Lenoel, C.,** Production par traitment mutagene de lignees cellulaires d'*Acer pseudoplatanus* L. anergiees a l'auxine, *C.R. Acad. Sci.,* 265, 1803, 1967.

89. **Lindsey, K.,** Manipulation, by nutrient limitation, of the biosynthetic activity of immobilized cells of *Capsicum frutescens* cv. *annuum, Planta,* 165, 126, 1985.

90. **Lindsey, K. and Yeoman, M. M.,** The synthetic potential of immobilized cells of *Capsicum frutescens* cv. *annuum, Planta,* 162, 495, 1984.

91. **Mader, M., Munch, P., and Bopp, M.,** Regulation of peroxidase patterns during shoot differentiation in callus cultures of *Nicotiana tabacum, Planta,* 123, 257, 1975.

92. **Malawer, C. L. and Phillips, R.,** Cell cycle in relation to induced xylem differentiation: tritiated thymidine incorporation in cultured tuber explants of *Helianthus tuberosus, Plant Sci. Lett.,* 15, 47, 1979.

93. **Mantell, S. H. and Smith, H.,** Cultural factors that influence secondary metabolite accumulations in plant cell and tissue cultures, in *Plant Biotechnology,* Mantell, S. H. and Smith, H., Eds., Cambridge University Press, Cambridge, 1983, 67.

94. **Masuda, H., Fukuda, H., and Komamine, A.,** Changes in peroxidase isoenzyme patterns during tracheary element differentiation in a culture of single cells isolated from the mesophyll of *Zinnia elegans, Z. Pflanzenphysiol.,* 112, 417, 1983.

95. **Matsumoto, T., Kanno, N., Ikeda, T., Obi, Y., Kisaki, T., and Nougchi, M.,** Selection of cultured tobacco cell strains producing high levels of Ubiquinone-10 by cell cloning technique, *Agric. Biol. Chem.,* 45, 1627, 1981.

96. **McHale, N. A.,** Conditions for strict autotrophic cultures of tobacco callus, *Plant Physiol.,* 77, 240, 1985.

97. **McNeil, M., Darvill, A. G., Frey, S. C., and Alberscheim, P.,** Structure and function of the primary cell walls of Plants, *Ann. Rev. Biochem.,* 53, 62, 1984.

98. **Mehta, A. R. and Fadia, V. P.,** Experimental induction of vascular differentiation in plant callus tissues, in *Seminar on Plant Cell, Tissue Organ Culture,* Johri, B. M., Ed., University of Delhi, India, 1967, 51.

99. **Meins, F.,** Mechanisms underlying the persistence of tumour autonomy in crown gall disease, in *Tissue Culture and Plant Science,* Street, H. E., Ed., Academic Press, New York, 1974, 233.

100. **Meins, F.,** Heritable variation in plant cell culture, *Annu. Rev. Plant Physiol.,* 34, 327, 1983.

101. **Meins, F. and Foster, R.,** Reversible cell heritable changes during the development of tobacco pith tissues, *Dev. Biol.,* 108, 1, 1985.

102. **Meins, F. and Lutz, J.,** The induction of cytokinin habituation in primary pith explants of tobacco, *Planta,* 149, 402, 1980.

103. **Meins, F., Lutz, J., and Foster, R.,** Factors influencing the incidence of habituation for cytokinin of tobacco pith tissue in culture, *Planta,* 150, 264, 1980.

104. **Miller, A. R., Crawford, D. L., and Roberts, L. W.,** Lignification and xylogenesis in *Lactuca* pith explants cultured in vitro in the presence of auxin and cytokinin: a role for endogenous ethylene, *J. Exp. Bot.,* 36, 110, 1985.

105. **Miller, A. R. and Roberts, L. W.,** Regulation of tracheary element differentiation by exogenous L-Methionine in callus of soybean cultivars, *Ann. Bot.,* 50, 111, 1982.

106. **Miller, A. R. and Roberts, L. W.,** Ethylene biosynthesis in xylogenesis in *Lactuca* pith explants cultured in vitro, in the presence of auxin and cytokinin: the effects of ethylene precursors and inhibitors, *J. Expt. Bot.,* 35, 691, 1984.

107. **Miller, A. R., Pengelley, W. L., and Roberts, L. W.,** Induction of xylem differentiation in *Lactuca* by ethylene, *Plant Physiol,* 75, 1165, 1984.

108. **Morel, G.,** Methode d'essai en serre des produits de lutte contre le mildiou la vigne, *Ann. Epiphyt. (Ser. Pathol. Veg.),* 13, 57, 1947.

109. **Morel, G.,** The effect of growth regulators on nitrogen metabolism of plant tissues, in *Proc. Int. Conf. Plant Tissue Culture,* White, P. R. and Grove, A. R., Eds., McCutchan Publishing, Berkeley, 1965, 93.

110. **Morel, G.,** Deviations du metabolisine ezote des tissus de crown gall, in *Les Cultures de Tissus de Plantes, (Colloq. Int. C.N.R.S.),* No. 193, 1971, 463.

111. **Morris, P. and Fowler, M. W.,** Growth and alkaloid content of cell suspension cultures of *Papaver somniferum, Planta Med.,* 39, 284, 1980.

112. **Mosebach, K. and Mosebach, R.,** Entrapment of enzymes and microorganisms in synthetic crosslinked polymers and their application in column techniques, *Acta Chem. Scand.,* 20, 2807, 1966.

113. **Muhitch, M. J. and Fletcher, J. S.,** Influence of culture age and supermidine treatment on the accumulation of phenolic compounds in suspension cultures, *Plant Physiol.,* 78, 25, 1985.

114. **Murillo, E. and de Jimenez, E. S.,** Glutamate synthase in greening callus of *Bouvardia ternifolia, Planta,* 163, 448, 1985.

115. **Näf, U.,** Studies on tumuor formation in *Nicotiana* hybrids. The classification of the parents into two etiological significant groups, *Growth,* 21, 167, 1958.

116. **Nato, A., Mathieu, Y., and Brangeon, J.,** Heterotrophic tobacco cell cultures during greening. II. Physiological and biochemical aspects, *Physiol. Plant.,* 53, 335, 1981.

117. **Nichio, M., Shoji, S., Ishii, T., Furuya, T., and Syono, K.,** Mass fragmentographic determination of indole-3-acetic acid in callus tissue of panax ginseng and *Nicotiana tabacum. Chem. Pharm. Bull.,* 24, 2038, 1976.

118. **O'Brien, T. P.,** Primary vascular tissues, in *Dynamic Aspects of Plant Ultrastructure,* Robards, A. W., Ed., McGraw-Hill, New York, 1974, 414.

119. **Pearson, D. W.,** Nicotine Production by Tobacco Tissue Cultures, Ph.D. thesis, Nottingham University, 1978.

120. **Peel, E.,** Photoautotrophic growth of suspension culture of *Asparagus officinalis* cells in turbidostats, *Plant Sci. Lett.* 24, 147, 1982.

121. **Petit, A., Delhaye, S., Tempe, J., and Morel, G.,** En evidence d'une relation destissus de crown gall. Mise biochimique specific entre les souches *Agrobacterium tumefaciens* et les tumero qu'elles induisant, *Ann. Physiol. Veg.,* 8, 205, 1970.

122. **Philips, R.,** Direct differentiation of tracheary elements in cultured explants by gamma-irradiated tubers of *Helianthus tuberosus, Planta,* 153, 262, 1981.

123. **Philips, R. and Hawkins, S. W.,** Characteristics of the inhibition of induced tracheary element differentiation by 3-aminobenzamide and related compounds, *J. Exp. Bot.,* 36, 119, 1985.

124. **Puschman, R., Ke, D., and Romani, R.,** Ethylene production by suspension-cultured pear fruit cells as related to their senescence. *Plant Physiol.,* 79, 973, 1985.

125. **Reinhard, E. and Alfermann, A. W.,** Biotransformation by plant cell cultures, *Adv. Biochem. Eng.,* 16, 49, 1980.

126. **Reinhard, E., Corduan, G., and Volks, O. H.,** Tissue culture of *Ruta graveolens, Planta Med.,* 16, 9, 1968.

127. **Richter, G.,** Blue light control of the level of two plastid mRNAs in cultured plant cells, *Plant Mol. Biol.,* 3, 271, 1984.

128. **Richter, G., Beckmann, J., Groβ, M., Hundrieser, M., and Schneider, J.,** Blue light-induced synthesis of chlorolast proteins in cultured plant cells, *Prog. Clin. Biol. Res.,* 102B, 267, 1982.

129. **Rier, J. P. and Beslow, D. T.,** Sucrose concentration and the differentiation of xylem in callus, *Bot. Gaz.,* 128, 73, 1967.

130. **Roberts, L. W.,** *Cytodifferentiation in Plants: Xylogenesis as a Model System,* Cambridge University Press, Cambridge, 1976.

131. **Rockem, J. S., Schwarzberg, J., and Goldberg, I.,** Autoclaved fungal mycelia increase diosgenin production in cell suspension cultures of *Dioscorea deltoidea, Plant Cell Rep.,* 3, 159, 1984.

132. **Roper, W., Schulz, M., Chaouiche, E., and Meloh, K. A.,** Nicotine production by tissue cultures of tobacco as influenced by various culture parameters, *J. Plant Physiol.,* 118, 463, 1985.

133. **Rubery, P. H. and Fosket, D. E.,** Changes in phenylalanine ammonia-lyase activity during xylem differentiation in *Coleus* and soybean, *Planta*, 87, 54, 1969.

134. **Sacristan, M. D. and Wendt-Gallitelli, M. F.,** Transformation to auxin autotrophy and its reversibility in a mutant line of *Crepis capillaris* callus culture, *Mol. Gen. Genet.*, 110, 355, 1977.

135. **Seeni, S. and Gnanam, A.,** Photosynthesis in cell suspension cultures of CAM plant *Chamaecereus sylvestrii* (Cactaceae), *Physiol. Plant.*, 49, 463, 1980.

136. **Seeni, S. and Gnanam, A.,** Relationship between chlorophyll concentration and photosynthetic potential in callus cells, *Plant Cell Physiol.*, 22, 1131, 1981.

137. **Seeni, S. and Gnanam, A.,** Growth of photoheterotrophic cells of peanut *Arachis hypogea* in still nutrient medium, *Plant Physiol.*, 70, 816, 1982.

138. **Seeni, S. and Gnanam, A.,** Photosynthesis in cell suspension culture of C_4 plant *Gisekia pharnacloides, Plant Cell Physiol.*, 24, 1033, 1983.

139. **Seyer, P. and Lescure, A. M.,** Evidence for changes in plastid mRNA populations during cytokinin-induced chloroplast differentiation in tobacco cell suspensions, *Plant Sci. Lett.*, 36, 59, 1984.

140. **Seyer, P. Marti, D., Lescure, A. M., and Peaud-Lenoel, C.,** Effect of cytokinin on chloroplast differentiation in cultured tobacco cells, *Cell Differ.*, 4, 187, 1975.

141. **Shininger, T. L.,** Is DNA synthesis required for the induction of differentiation in quiescent root cortical parenchyma, *Dev. Biol.*, 45, 137, 1975.

142. **Smith, H. H.,** Plant genetic tumors, *Prog. Exp. Tumor Res.*, 15, 138, 1972.

143. **Sogeke, A. K. and Butcher, D. N.,** The effect of inorganic nutrients on the hormonal requirements of normal, habituated and crown gall tissue cultures, *J. Exp. Bot.*, 27, 785, 1976.

144. **Sunderland, N.,** Pigmented plant tissues in culture. I. Auxin and pigmentation in chlorophyllous tissues, *Ann. Bot.*, 30, 253, 1966.

145. **Sung, Z. R.,** Relationships of indole-3-acetic acid and tryptophan concentrations in normal and 5-methyl tryptophan resistant cell lines of wild carrots, *Planta*, 145, 339, 1979.

146. **Syono, K.,** Correlation between induction of auxin-non-requiring tobacco calluses and increase in inhibitor/s of IAA-destruction activity, *Plant Cell Physiol.*, 20, 29, 1979.

147. **Syono, K. and Furuya, T.,** Effects of cytokinin on the auxin requirement and auxin contents of tobacco callus, *Plant Cell Physiol.*, 13, 843, 1972.

148. **Tanaka, H. and Fugimori, T.,** Accumulation of phytuberin and phytuberol in tobacco callus inoculated with *Pseudomonas solanacearum* or *P. syringae, Phytochemistry*, 24, 1193, 1985.

149. **Teuscher, E.,** Problems der Produktion sekundarer Pflanzenstoße mit Hilfe von Zellkulture, *Pharmazie*, 28, 6, 1973.

150. **Thomas, E. and Street, H. E.,** Organogenesis in cell suspension cultures of *Atropa belladonna* and *A. belladonn* cv. Lutea, *Ann. Bot.*, 34, 657, 1970.

151. **Tiburcio, A. F., Ingersoll, R., and Galston, A. W.,** Modified alkaloid pattern in developing tobacco callus, *Plant Sci.*, 38, 207, 1985.

152. **Tietjen, K. G., Hunkler, D., and Matern, U.,** Differential response of cultured parsley cells to elicitors from two nonpathogenic strains of fungi. I. Identification of induced products as coumarin derivatives, *Eur. J. Biochem.*, 131, 401, 1983.

153. **Torrey, J. G.,** Hormonal control of cytodifferentiation in agar and cell suspension cultures, in *Biochemistry and Physiology of Plant Growth Substances*, Wightman, F. and Setterfield, G., Eds., Runge Press, Ottawa, Canada, 1968, 843.

154. **Torrey, J. G.,** Cytodifferentiation in plant cell and tissue culture, in *Les cultures de Tissus de Plantes*, Colloq. Int. CNRS Paris No.193, 177, 1971.

155. **Torrey, J. G.,** Tracheary element formation from single cells in culture, *Physiol. Plant.*, 35, 158, 1975.

156. **Torrey, J. G., Fosket, D. E., and Hapler, P. K.,** Xylem formation: a paradigm of cytodifferentiation in higher plants, *Am. Sci.*, 59, 338, 1971.

157. **Tso, T. C., Burk, L. G., Dieterman, L. J., and Wender, S. H.,** Scopoletin, scopolin and chlorogenic acid in tumours of interspecific *Nicotiana* hybrids, *Nature (London)*, 204, 779, 1964.

158. **Tsuzuki, M., Miyachi, S., Sato, F., and Yamad, Y.,** Photosynthetic characteristics and carbonic anhydrase activity in cells cultured photoautotrophically and mixotrophically and cells isolated from leaves, *Plant Cell Physiol.*, 22, 51, 1981.

159. **Turgeon, R.,** Differentiation of wound vessel members without DNA synthesis, mitosis or cell division, *Nature (London)*, 257, 806, 1975.

160. **Turgeon, R.,** Cytokinesis, cell expansion, and the potential for cytokinin autonomous growth in tobacco pith, *Plant Physiol.*, 70, 1071, 1982.

161. **van Wezel, A. L.,** The large scale cultivation of diploid cell strains in microcarrier culture. Improvement of microcarriers, *Dev. Biol. Stand.*, 37, 143, 1967.

162. **van Wezel, A. L.,** Growth of cell strains and primary cells of microcarriers in homogeneous culture, *Nature (London)*, 216, 64, 1967.

163. **Vyskot, B., Karpfel, Z., and Bezdek, M.,** Hormonal control of 5-Bromodeoxyuridine (BuDR) tolerance in crown gall tumours and habituated tissues, *Planta,* 137, 247, 1977.

164. **Vyskot, B. and Novak, F. J.,** Habituation and organogenesis in callus cultures of chlorophyll mutants of *Nicotiana tabacum* L., *Z. Pflanzenphysiol.,* 81, 34, 1976.

165. **Watson, B. and Halperin, W.,** Reinvestigation of the effect of hormone and sugars on xylogenesis in cultured Jerusalem artichoke *(Helianthus tuberosus)* tuber slices with particular emphasis on the effects of different methods of media preparation and tissue analysis, *Z. Pflanzenphysiol.,* 101, 145, 1981.

166. **West, C. A.,** Fungal elicitors of the phytoalexin response in higher plants, *Naturwissenschaften,* 68, 447, 1981.

167. **Wetmore, R. H. and Rier, J. P.,** Experimental induction of vascular tissues in callus of angiosperms, *Am. J. Bot.,* 50, 418, 1963.

168. **Wetmore, R. H. and Sorokin, S.,** On the differentiation of xylem, *J. Arnold Arbor.,* 36, 305, 1965.

169. **White, P. R.,** Potentially unlimited growth of excised plant callus in an artificial medium, *Am. J. Bot.,* 26, 59, 1939.

170. **Widholm, J. M.,** Relation between auxin autotrophy and tryptophan accumulation in cultured plant cells, *Planta,* 134, 103, 1977.

171. **Widholm, J. M.,** Selection and characterization of amino acid analog resistant plant cell cultures, *Crop Sci.,* 17, 597, 1977.

172. **Wijnsma, R., Go, J. T. K. A., Weerden, I. N. V., Harkes, P. A. A., Verpoorte, R., and Sverdsen, A. B.,** Anthraquinones as phytoalexins in cell and tissue cultures of *Cinchona* sp., *Plant Cell Rep.,* 4, 241, 1985.

173. **Wilson, G. and Balague, C.,** Biosynthesis of anthraquinones by cells of *Galium mullugo,* grown in a chemostat with limiting sucrose or phosphate, *J. Exp. Bot.,* 36, 485, 1985.

174. **Wilson, J. W., Roberts, L. W., Gresshoff, P. M., and Dircks, S. J.,** Tracheary element differentiation induced in isolated cylinders of lettuce pith: a bipolar gradient technique, *Ann. Bot.,* 50, 605, 1982.

175. **Wink, M.,** Composition of the spent cell culture medium. I. Time course of ethanol formation and the excretion of hydrolytic enzymes into the medium of suspension cultured cells of *Lupinus polyphyllus, J. Plant Physiol.,* 121, 287, 1985.

176. **Yamada, Y., Imaizumi, K., Sato, F., and Yasuda, T.,** Photoautotrophic and photomixotrophic culture of green tobacco cells in a jar fermentor, *Plant Cell Physiol.,* 22, 917, 1981.

177. **Yamada, Y. and Sato, F.,** Photoautotrophic culture of chlorophylous cultured cells, *Plant Cell Physiol.,* 19, 691, 1978.

178. **Yamada, Y. and Sato, F.,** Selection for photoautotrophic cells, in *Handbook of Plant Cell Culture,* Evans, D. A., Sharp, W. R., Ammirato, P. V., and Yamada, Y., Eds., Macmillan, New York, 1983, 489.

179. **Yasuda, T., Hashimoto, T., Sato, F., and Yamada, Y.,** An efficient method of selecting photoautotrophic cells from cultured heterogenous cells, *Plant Cell Physiol.,* 21, 929, 1980.

180. **Yeoman, M. M., Miedzybrodzka, M. B., Lindsey, K., and Mchauglan, W. R.,** The synthetic potential of cultured plant cells, in *Plant Cell Cultures: Results and Perspectives,* Sala, F., Parisi, B., Cella, R., and Ciferri, O., Eds., Elsevier, Amsterdam, 1980, 327.

181. **Zenk, M. H., Shagi, H., Arens, H., Stokigt, J., Weiler, E. W., and Deus, B.,** Formation of indolealkaloids, serpentine, and ajmalacine in cell suspension cultures of *Catharanthus roseus,* in *Plant Tissue Culture and Its Biotechnological Applications,* Barz, W., Reinhard, E., and Zenk, M. H., Eds., Springer-Verlag, Berlin, 1977, 27.

Chapter 3

CELL TOTIPOTENCY

I. INTRODUCTION

Vegetative or clonal propagation of plants is an ancient practice for crop improvement. In clonal propagation it is possible to have genetically identical plants compared to heterogenity seen in populations raised from seeds. The basis of vegetative or clonal propagation lies in the regenerative capacity of plant cells. An extension of the regenerative process to a wide range of cells is possible by the technique of tissue and cell culture. Employing this technique, it has been possible to have a large number of plants in a relatively short time and space. According to a conservative estimate, a million-fold increase in the number of plants per year is possible through tissue culture, over the conventional means.[277] The propagation of plants through tissue or cell culture has been described as "micropropagation".

Plant cells, in addition to their utility in micropropagation, provide new avenues for crop improvement. Genetic techniques can be applied directly to plant cells. However, a prerequisite for the application of genetic techniques to cells for crop improvement is the devise of an efficient protocol for regeneration of plants from cells. This is the basic theme of this chapter.

At present, cells from a large number of plants have been shown to readily regenerate into entire plants, but cells from many other plants, in particular, cereals and seed legumes, have been found to be recalcitrant. To overcome this limitation it is essential to have an understanding of the process of regeneration. Also, not all cells from any part regenerate to produce plants. This raises the question of competence of certain cells for regeneration. Further, cells after remaining in culture undergo undesired modifications and are no longer capable of regeneration; in other words, they progressively lose the potential for regeneration. These questions are also addressed in this chapter.

II. CONCEPT

Plant cells, unlike animal cells, even after undergoing differentiation *in situ*, retain the capacity to revert to the meristematic state. This reversion, leading to an unorganized growth of cells, is described as dedifferentiation. In addition, a plant cell has an inherent capacity to regenerate into an entire plant. This capacity is described as totipotency.

The earliest experimental evidence in support of the concept of totipotency can be seen in experiments describing the formation of plants from plant parts which were dissected into smaller and smaller fragments.[391] However, an unequivocal evidence of totipotency of higher plant cells was possible when unorganized tissues from tobacco hybrid[410] *(N. glauca × N. langsdorffii)* and carrot[239,240] were found to differentiate into shoots and roots, respectively. Subsequent work has confirmed that tissues from numerous other plants can undergo dedifferentiation to form a callus, and redifferentiate to form roots and shoots.

As for the concept of totipotency, the most illuminating work is on tobacco tissue which provided the first insight into the control of differentiation. From a callus tissue, at will, it was possible to have either roots or shoots, or both.[308] Following this, the demonstration of formation of embryos from callus tissue of carrot,[274,275] and cell suspension of carrot,[327] and finally, the demonstration of plant formation in culture initiated from a single cell of tobacco[386] and embryo formation in culture intiated from a single cell of carrot[10] led to the firm establishment of the concept of totipotency in plants.

At the outset, a clarification about the concept of cell totipotency is in order. According

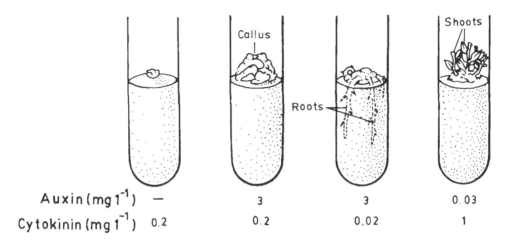

| Auxin (mg l^{-1}) | — | 3 | 3 | 0.03 |
| Cytokinin (mg l^{-1}) | 0.2 | 0.2 | 0.02 | 1 |

FIGURE 1. Diagrammatic representation of auxin-cytokinin interaction in organogenesis of a callus culture.

to this concept, although each cell is considered to be totipotent, it does not mean that an individual cell regenerates directly into a plant. Instead, on culture, an individual cell undergoes proliferation to produce a mass of cells and from this tissue in turn arise roots and shoots or embryos.

III. CONTROL OF REGENERATION

Initial reports on formation of shoot-buds on tobacco[410] tissue on submersion in liquid medium and root formation from carrot tissue[240] were short of explanations about the control of regeneration. An analysis was, however, possible when auxin was shown to stimulate root formation and inhibit shoot formation in tobacco tissue.[306] Further, a purine,[309] adenine, was found to promote shoot formation and inhibit the effect of auxin. These results were extended to other plants such as *Helianthus*[189] and *Ulmus*.[142] The extent of organogenesis was dependent on the concentration or proportion of adenine and auxin.[310] However, more significant was the discovery of kinetin,[213] a purine derivative, later renamed as cytokinin. The kinetin or cytokinin is more potent than adenine and its use for tissue organization led to the classical formulation that organogenesis in a tissue is the function of a balance between auxin and cytokinin.[308] Both these substances are essential for undifferentiated growth of tobacco tissue, but a high level of cytokinin to auxin ratio favors shoot formation and the reverse is true for root formation (Figure 1). At certain levels it is possible to have both shoots and roots or none, i.e., continued unorganized state. These results formed the basis for regulation of regeneration in plants and negated the earlier concept[401] that there are specific organ-forming substances: rhizocaline for root and caulocaline for stem or shoot.

IV. PATTERN FORMATION AND FACTORS CONTROLLING REGENERATION

Organization in an unorganized tissue can be viewed in different perspectives. The cells and tissues in cultures are capable of organization to form shoots, roots, embryos, and even flowers. The differentiation is *de novo*, it is not from preexisting primordia. Each of these differentiating organs conforms to a structural pattern which is the manifestation of the preceding physiological, biochemical, and biophysical events. A good deal of information is available about factors controlling organization, but its final analysis, the regulation of regeneration at the molecular level, remains to be resolved.

A. Formation of Monopolar Structure (Organogenesis)

Control of organogenesis, formation of roots or shoots, was demonstrated for the first time in tobacco tissue[308] and since then, this has become a model system and much of the information has come from work on tobacco tissue.

A callus tissue remains unorganized on callus maintenance medium and essentially, it comprises parenchymatous cells, each of which is highly vacuolated with a sparse cytoplasm and diffuse peripherally placed nucleus. Some sort of cell differentiation can be seen in a callus tissue, in the form of scattered tracheidal elements. An unorganized growth of a callus tissue is due to random divisions. However, on transfer of callus to differentiation medium, within a period of 2 to 8 days can be seen centers of mitotic activity. Continued divisions in these centers result in the formation of meristematic zones or meristemoids, which are generally located on the surface of callus, but are also embedded in the tissue. In tobacco tissue maintained on agar medium these meristemoids are preferentially on the lower half. A meristemoid[222,369,373] originates either from a single cell or a group of cells and comprises small isodiametric cells, each of which has small vacuoles, a dense cytoplasm, and a large nucleus. Another feature of these meristemoid cells is their strong reaction towards stains for RNA and protein.[369,370] In addition, each cell of the meristemoid is rich in organelles and its prominent nucleus has a prominent nucleolus.

Initially, a meristemoid is plastic, capable of giving rise to either a root or a shoot primordium.[29] These meristemoids, to begin with, also appear to be apolar, but directional division activity in them leads to the formation of specific organ primordia.[360] Plasticity and polarity of the meristemoids on a callus needs to be looked into in greater detail.

An organ primordium for either a root or a shoot is basically a unipolar structure in continuation with the main tissue. Vascular elements in this primordium appear at a later stage and there may or may not be any continuity between them and tracheidal elements of the callus tissue.

It is of interest to know how highly vacuolated cells of callus give rise to meristematic cells. A differentiating tissue of tobacco accumulates starch[369,370] and paracrystalline bodies.[287] During meristemoid formation, starch and vacuolar contents progressively decrease and finally, the starch and paracrystalline bodies are also no longer visible.

1. Factors Affecting Organogenesis

The factors affecting organogenesis in an unorganized tissue can be broadly classified into chemical and physical environments. In determining the exact role of any specific variable in control of organogenesis, the major difficulties encountered are lack of knowledge about: (1) correlation between chemical and physical environments, (2) interaction between endogenous and exogenously supplied chemicals, and (3) mode of action of chemical substances affecting organogenesis. Therefore, no generalization is possible and the conclusions arrived at have to be tentative. Despite these difficulties, more and more information is being added which will help unravel the control of organogenesis in particular and differentiation in general.

a. Chemical Environment

Since the early work on tobacco tissue,[308] it is evident that expression of organogenesis can be controlled by an external supply of chemicals. High auxin and a low cytokinin level promotes continuous proliferation of cells — unorganized growth. By contrast, a high cytokinin and low auxin level favors shoot-bud formation. Further, the formation of specific organ, root, or shoot is a function of quantitative interaction between auxin and cytokinin. Basically, an auxin promotes the formation of roots[105,306] and a cytokinin favors shoot-bud formation.[308] The shoot-forming effect of cytokinin is modified by auxin. As low as $5\mu M$ of IAA is sufficient to inhibit spontaneous shoot-bud formation in tobacco tissue, and about

15,000 molecules of adenine or 2 molecules of kinetin are required to counteract 1 molecule of IAA.[307] Auxin inhibition of shoot-bud formation is also supported by the finding that inclusion of antiauxin, N-1-napthylpthalamic acid,[90] or auxin transport inhibitors, DPX-1840 and chloroflurenol,[52] promoted shoot-bud formation in tobacco tissue.

Auxin-cytokinin balance has been found to regulate organogenesis in a large number of plants, particularly members of Solanaceae, however, there are exceptions. In tuber tissue of *Cyclamen*, adenine promoted the number of regenerants, but the pattern of organogenesis, root, or shoot was determined by concentration of auxin, NAA.[208,329] Similarly, callus tissue of *Armoracea*[291] formed shoot-buds in the presence of auxin, and cytokinin suppressed shoot-bud formation whereas in seedling tissue of *Lactuca*,[74] cytokinin was effective in the presence of a low level of auxin and in cultures, the explants from cauliflower head[207] formed shoot-buds only in the presence of auxin. In this context is also the finding that in callus tissue of *Medicago*[392] a high level of auxin and a low level of cytokinin resulted in shoot formation while the reverse was true for root formation. Also, in bulbous plants, such as hyacinth,[256] *Ornithogalun thyrsoides*,[137] *Lilium speciosum*,[382] and *Urginea indica*,[147] adventitious bud formation from storage tissue was stimulated by an exogenous supply of auxin. In this series of exceptions is also the instance of root formation on removal of auxin from the medium in *Atropa*,[356] and this was stimulated by inclusion of an antiauxin (tropic acid, 1-naphthoxy-acetic acid). However, these findings should not be taken to be anomalous unless a determination of the levels of endogenous hormones is done and an interaction with exogenous hormones worked out.

Auxin is a root-forming substance and is also essential for induction and maintenance of a callus. Since natural auxin, IAA, is sensitive to degradation by oxidation,[182] synthetic auxins, NAA, IBA, and 2,4-D, are preferred. However, in tobacco tissue, of the various indole derivatives, none proved superior to IAA.[360] At higher concentration, 2,4-D promotes cytological instability and this is accompanied by loss of morphogenic ability. Nevertheless, 2,4-D has been the auxin of choice for most of the graminaceous species worked out. This has to be viewed in the light of finding that in monocot tissues 2,4-D is metabolized into a physiologically inactive glycoside derivative whereas in dicots, this is replaced by physiologically active amino acid conjugates.[91,92]

For shoot formation, of the various cytokinins — 6-furfurylaminopurine (kinetin), 6-benzylaminopurine (BAP), 6-tetrahydropyrane adenine (SD-8339), 6-7-7-dimethylallylamino purine (2iP), and triacanthene — BAP has proved to be most effective. As for cytokinin, it is worth recalling that its discovery was an outcome of earlier work which indicated the role of adenine in shoot formation.[309] Of the other purines, guanine[310] was also effective. Cytokinin effect on shoot formation of tobacco tissue is also simulated by urea derivatives[140] and chelating agents.[173] The mode of cytokinin action in induction of shoot-buds remains to be worked out. More recently, an indirect evidence indicates the involvement of calcium and calmodulin.[349]

Gibberellins tend to suppress organogenesis in tobacco tissue[222] and this repression is not reversed by gibberellin antagonists, AMO 1618, CCC, and phosphon-D or B-995, which also inhibit organogenesis.[223] The repression of organogenesis could be overcome by including excess levels of auxin and cytokinin. However, gibberellin stimulates shoot formation in callus tissues of *Chrysanthemum*,[80] *Arabidopsis*,[235] and rooting in *Phaseolus*[121] callus. Repression of organogenesis by gibberellins has been explained in terms of excess of endogenous gibberellins in the tissue. The same reasoning can also explain gibberellin-induced stimulation of shoot formation in *Chrysanthemum* and *Arabidopsis*, where endogenous levels are likely to be suboptimal. It can also be extended to account for abscisic acid (ABA) - induced shoot-bud formation in sweet potato,[417] where possibly supra optimal level of gibberellin is countered by ABA. In fact, repression of organogenesis by gibberellin in tobacco tissue[365] was overcome by a very low level (10^{-6} *M*) of ABA.

Many of auxin effects, aside from promotion of cell division, are considered to be mediated via auxin control of ethylene synthesis.[46] It has also been suggested that gradients of auxin and ethylene in cells may account for control of rate, extent, and orientation of cell growth,[246] the basic changes which occur during cell differentiation and have eluded elucidation. An exogenous supply of ethylene to tobacco tissue during the first 5 days of differentiation inhibited organogenesis, but when given during 5 to 10 days it speeded up primordium formation.[139] Also shoot-forming tobacco tissue produced less endogenous ethylene than nonshoot-forming tissue, indicating that ethylene is antagonistic to organogenesis. However, the role of ethylene in organogenesis needs to be studied in detail because a low concentration of ethylene is implied in modulating normal plant growth and development. *Digitalis obscura* is endowed with high regenerative potential and plants are possible from almost every part. The hypocotyl explants of this plant formed roots in the presence of ethylene and shoot formation was possible in presence of IAA and kinetin. Addition of ethylene to this medium promoted shoot formation.[252] Also, the promotory effects of NAA, TIBA, tissue wounding, and temperature in shoot-bud formation of *Lilium*[383] can be explained through ethylene biosynthesis. This was confirmed when plantlet regeneration was inhibited by amino-ethoxy-vinyl-glycine (AVG), an inhibitor of ethylene biosynthesis. The effect of AVG was counteracted by 1-aminocyclopropane-1-carboxylic acid (ACC) and TIBA.[384] The synthesis of ethylene was also demonstrated and bud formation by the tissue was proportional to its ethylene production.[385]

b. Physical Environment

A number of variables in the physical environment also affect organogenesis. Unfortunately, it is difficult to correlate them with the chemical environment. Although the first report on shoot-bud formation[410] from tissue of tobacco hybrid (*N. glauca* × *N. langsdorffii*) on submersion of callus in liquid medium was confirmed,[306] extensive bud formation on tobacco tissue occurs on agar medium compared to a very low response in liquid medium.[75]

Under various light regimes organogenesis is possible from different tissues. This indicates that a photoperiodic requirement is not a critical determinant.[224] Initiation of organogenesis on tobacco tissue is also possible in dark[370] as well as in continuous light. However, callus of *Pelargonium hortorum*[257] formed shoot-buds in alternating light and dark periods and remained undifferentiated in continuous light.

As for quality of light, blue light was found to be important for shoot-bud differentiation by tobacco tissue.[400] In a detailed analysis it was found that ultraviolet light (371 nm) inhibited organogenesis at high intensity in tissues of tobacco[96,296] and onion.[96] However, at low intensity it was stimulatory.[296] Red and far-red lights were ineffective and blue light supported organogenesis in tobacco tissue. An indication of phytochrome involvement in shoot-bud formation is seen in cotyledon cultures of lettuce[152] and embryo callus of pine.[153] This area of research deserves more attention and is likely to be rewarding. In order to emphasize, it will not be out of place to include examples of regeneration from intact tissues. Initiation of lateral roots on isolated pea roots is promoted by far-red (FR) light and inhibited by red light.[102] Differentiation of adventitious roots within 12 to 24 hr on the hypocotyl sections of mustard seedlings is also a phytochrome-mediated response. Operationally, continuous FR light induces *de novo* formation of roots. Growth regulators IAA, GA_3, kinetin, ABA, and ethylene were ineffective and were rather inhibitory indicating that hormones are no causal links in this phytochrome-mediated response.[254] Shoot-bud differentiation in callus cultures of *Brassica oleracea* also seems to be a phytochrome-mediated response.[12]

The effect of temperature on organogenesis by callus cultures has not been investigated in detail, save for the sole report of callus tissue of tobacco.[306] Interestingly, 18°C was found to be optimal for shoot initiation with 12 and 33°C being the extreme limits. This is contrary to 25°C, the normal temperature employed for maintenance of cultures.

The role of osmotic potential in organogenesis was introduced by work on callus tissue of potato.[301] The tissue exhibited greening prior to shoot formation on a medium of 200 to 400 mmol which was possible by inclusion of 0.2 to 0.3 M mannitol. This is confirmed and it is shown that optimal conditions for shoot-bud formation are quite different from root formation.[166] Shoot-buds were produced when potato tuber tissue was cultured on medium containing 0.25 M mannitol with zeatin and indole-acetic acid under relatively high irradiation while roots were formed when medium contained 29 mM sucrose, kinetin, IAA, and was exposed to dark.

2. Organogenesis — An Overview

A perusal of the above account indicates that in *N. tabacum*, the model system, it is easy to control root/shoot morphogenesis by varying the chemical environment. However, this plasticity can be a constraint for further research, particularly for an understanding of the regulation of morphogenesis. In order to define the problem critically it is necessary to have mutants of morphogenesis or compare within the same genus two genotypes, one which can readily regenerate, i.e., *N. tabacum*, and the other, either a reluctant cultivar or a hybrid. Also it has to be emphasized that control of organogenesis remains to be achieved in many of our crop plants, particularly cereals[63] and seed legumes. Tissue and cell suspension cultures raised from most of our seed legumes and cereals readily regenerate to form roots, but are recalcitrant to shoot formation.

An understanding of the process of regeneration will be possible by the induction, recovery, and characterization of mutants defective at certain stages. To this effect, a limiting factor is a complex nature of form and function in higher plants. Therefore, an alternative approach is stage-specific inhibition (phenocritical time) in organogenesis.

a. Phenocritical Times in Organogenesis

Given a complete medium, there are discrete steps which precede morphological differentiation of shoot-bud formation. This was discovered when leaf explants of *Convolvulus* were cultured on shoot-inducing medium comprised of salts of MS medium, sucrose, vitamins, auxin (0.5 mg/ℓ IAA) and 2-isopentyladenine (7mg/ℓ). On this induction medium the process of shoot induction could be resolved into different stages (Figure 2) sensitive to inhibitors. This stage-specific inhibition reflects phenocritical times[57] in development rather than general metabolic toxicities.

The process includes a time when differentiation is sensitive to inhibition by salicylates, followed by a time sensitive to TIBA, which is followed in turn by a time sensitive to sorbitol and culminates in cells or group of cells determined for shoot formation. This process also includes a time sensitive to inhibition by ribose, although its place in the order of events is not fully established. There is also sensitivity to ammonium ions (or lack of nitrate) at or near the time the explant becomes determined for shoot formation. Similar study in other systems will help in the understanding of organogenesis.

B. Formation of Bipolar Structure (Embryogenesis)

Organization in a callus tissue is also possible by the formation of embryo-like structures. These structures, unlike root and shoot, are strictly bipolar with a distinct root and a cotyledonary pole. This type of structure has been variously described as embryo, embryoid, adventive embryo, somatic embryo, accessory embryo, supernumerary embryo, and even neomorph. The more common usage is the term "embryoid", but without a justification. A structure similar to an embryo in organization as well as function should be described as embryo.

Only that structure which has a distinct bipolar axis (Figure 3) should be described as an embryo.[117] An emphasis on this point is essential, because embryo formation is a less frequent

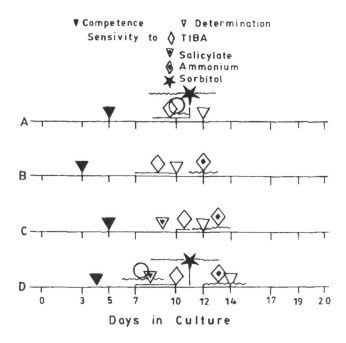

FIGURE 2. Phenocritical times in the process of in vitro shoot formation from leaf segments of *Convolvulus*. (From Christianson, M. L. and Warnick, D. A., *Dev. Biol.*, 101, 382, 1984. With permission.)

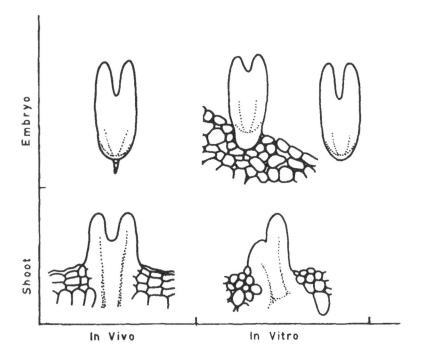

FIGURE 3. Diagrammatic representation of differences between shoot and embryo formed in vivo and in vitro. (Redrawn after Haccius, B., *Phytomorphology*, 28, 74, 1978. With permission.)

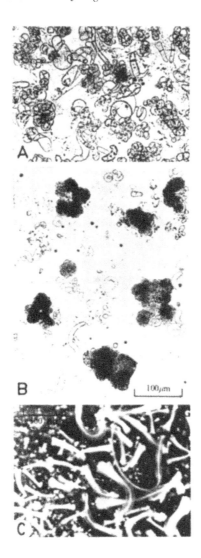

FIGURE 4.   Embryo formation in cell suspension cultures of *Daucus carota*. (A) Cell suspension culture showing embryogenic clumps and free-cells. (B) Embryogenic clumps magnified. (C) Free embryos in various stages of development. (From Smith, S. M. and Street, H. E., *Ann. Bot.*, 38, 223, 1974. With permission).

mode of differentiation in tissue cultures, and a mere bifurcation of shoot primordium should not be taken as the cotyledonary end of an embryo. The most distinctive feature of an embryo is its discrete and closed radicular end.[118] Other features also distinguish an organ, shoot or root, from an embryo. An organ is in organic union with the main tissue from which it has originated and can also have continuity of its vascular elements with that of callus tissue. Both these features are lacking in an embryo.

Since the discovery of embryo formation in tissue cultures of carrot,[274,275,327] it is shown to occur (Figure 4) in all other umbellifers worked out so far. It also occurs in widely different plants such as *Citrus* spp.,[169,265,266,290] *Ranunculus sceleratus*,[176] *Macleaya cor-*

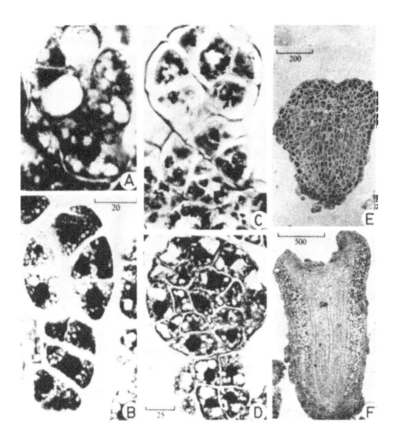

FIGURE 5.   Origin and development of embryos in cell suspension cultures of *Daucus carota*. (A) Four-celled proembryo. (B, C) Later stages of development. (D) Young embryo showing demarcation of an outer epidermal layer from central cells; at the base is also seen a suspensor-like structure. (E) Heart-shaped embryo. (F) Torpedo-stage embryo. (From McWilliam, A., Smith, S. M., and Street, H. E., *Ann. Bot.*,, 38, 243, 1974. With permission.)

data,[174] *Coffea camphora*,[322] and *Atropa belladonna*.[356] Among monocots, it occurs in grasses *Bromus inermis*[65] and *Lolium*,[70] millets *Pennisetum*,[388] *Panicum*,[196] and *Sorghum*,[403] and cereals *Triticum*[247] and *Zea*.[197] It is also shown in a number of legumes such as *Trifolium*[62,203] and *Medicago*.[155,199] Other important plants where embryo formation occurs are coffee,[316] cassava,[321] oil palm,[245] coconut palm,[36] cotton,[93] cacao,[180] walnut,[378] apple, and pear.[211,212]

1. Origin and Development of an Embryo

In its development, an embryo differentiating from a tissue simulates the sequences of a zygotic embryo such as proembryonal, globular, heart-shaped, and torpedo stages[274,275] (Figure 5).

The origin of an embryo has attracted considerable attention and has also been debated at length. On the basis of developmental sequence (almost identical patterns of cell division with that of zygotic embryos), it was proposed that embryos in carrot tissue arise from individual cells on the periphery of callus.[277] Also, on the basis of observation of free-floating embryos in suspension cultures of carrot it was postulated[325] that (1) isolated individual cells develop directly, without callus formation, to form proembryos, and these in turn develop into plants, and (2) isolation of an embryo-forming cell is a prerequisite for the initiation of an embryo. However, when isolated single cells of carrot were subjected to embryogenesis, to begin with, a multicellular aggregate was formed and on it in turn differentiated the embryos.[10] This clearly indicates that isolated single cells do not directly embark upon embryogenesis. Also, an electron microscopic study has clearly indicated that embryo-forming cells maintained cytoplasmic connections with the neighboring callus cells.[358] This is contrary to the assumption that physiological isolation of embryo-forming cells is a necessary prerequisite for embryo formation.[325]

It has also been clarified that embryos do not originate from ordinary vacuolated cells of callus, but arise from small meristematic cells which can be either individual cells or a group of such cells.[210] However, the origin of an embryo from a single cell remains to be demonstrated unequivocally.[263,371] The cells of carrot destined to form embryos are highly cytoplasmic, and each possess a large, diffuse nucleus. These cells stain intensely with protein and RNA stains.[210] Studies on another embryo-forming plant, *Ranunculus sceleratus,*[177,358] support these findings and also reveal that embryo formation is possible from cells which are deeper in callus.[177] A majority of embryos, however, originate from cells on the periphery of callus either directly from embryogenic clump/callus nodule or from a single peripheral cell. Multicellular origin of an embryo results in its contact/fusion with the parent tissue over a broad area in the principle root pole. By contrast, unicellular origin results in loose and freely dispersed embryo. This is, however, questioned.[119]

The process of embryo formation from cell clusters of carrot, when made to divide synchronously, is divisible into three phases:[100] the first phase (0 to 3 days), during which cells divide slowly, the second phase (3 to 4 days), during which cells divide very rapidly in a determined manner and lead to the formation of a globular embryo, and the third phase (4 to 6 days), during which, again, cell divisions occur at a slower rate. The second phase is considered to be a crucial phase of embryogenesis. At the end of this phase in a globular embryo could be distinguished three regions: a transparent region, destined to differentiate into shoot region, an opaque region, destined to differentiate into root, and another opaque region which does not grow. This suggests that determination of differentiation occurs by the end of the second phase. A similar study of other embryo-forming tissues will help in the characterization of processes. This will in turn help in the understanding of physiology of embryo formation by somatic cells.

Coordination within a group of cells leading to the differentiation of an embryo remains to be defined. Recent work[39] indicates that an ionic current directed along the developmental axis is a determining factor until the early torpedo stage.

Also of interest in this context is the geometric analysis[292] of embryo formation in carrot cell cultures from the postglobular stage. The globular stage is characterized by isodiametric cell expansion. The heart stage embryo is initiated by elongation of the future hypocotyl axis. It is before the emergence of cotyledons. This intermediate stage is termed by the authors as oblong embryo. The formation of torpedo embryo from a heart embryo does not involve any apparent change in growth of various parts, but results from a continuous increase in overall size.

2. Factors Affecting Embryogenesis

The factors affecting embryo formation can be classified into two categories, tissue environment and tissue origin. In evaluating the role of any specific factor, the difficulties remain the same as described for organogenesis.

a. Tissue Environment

In one of the earliest results on embryo formation from carrot tissue, coconut milk was one of the ingredients of the medium,[327] and consequently it was emphasized that coconut milk is essential for embryogenesis[324] and also serves to nourish the developing embryos. This was, however, an incorrect inference because it had already been demonstrated that embryogenesis in carrot is possible on a defined nutrient medium.[276]

The decisive step in the induction of embryos is the transfer of callus from auxin to auxin-free medium and increasing the nitrogen level of medium by including amino acids.[276] Therefore, embryo formation is considered to be a two-step process, and important factors controlling it are nitrogen and auxin level of medium.

i. Nitrogen

The pioneering work on embryo formation, employing tissue of domestic carrot, was done on White's nutrient medium,[409] which has a very low level of nitrogen (3.2 mM) in the form of nitrate only. Embryo formation on this tissue is possible in the presence of inorganic (nitrate or ammonium) as well organic forms of nitrogen, amino acids, and amides.[352] White's nutrient medium alone is not supportive of embryo formation, but it could be made to be so by inclusion of a high amount of KNO_3 (46 mM) or a low amount of nitrate (3.2 mM) along with a low amount of glutamine (5 mM).

By contrast, tissue from wild carrot formed embryos only when the medium contained some amount of reduced nitrogen.[125,126] The tissue initiated on medium with nitrate as the sole source of nitrogen failed to form embryos on auxin-free medium. Instead, a small addition of reduced nitrogen (5 mM NH_4Cl) along with 55 mM KNO_3 favored embryo development. The necessity of reduced nitrogen could be met even during callus proliferation. After the callus was raised on nitrate and ammonium medium, embryogenesis was possible in medium having only nitrate and lacking ammonium. Therefore, in wild carrot, unlike domestic, nitrate as a sole source of nitrogen does not support embryogenesis and it is possible on ammonium medium. However, when employing NH_4Cl as a sole source of nitrogen, a caution is necessary about the pH of the medium; it drops from 5.4 to 4 to 3.5 within 4 days, and a low pH is inhibitory to embryogenesis. Consequently, it is convenient to employ ammonium as well as nitrate.

This anomaly, the nonessentiality of the reduced form of nitrogen for embryogenesis by tissue of domestic carrot, appears resolved when the endogenous level of nitrogen in the tissue of this variety is accounted. In this variety, for embryogenesis to proceed, an endogenous level of nitrogen, around 5 mM of NH_4 per kilogram of fresh weight of tissue, is essential. This level, in the tissue, can be achieved[352] by supplying from outside either a low level of NH_4^+ (2.5 mM) or high level of NO_3^- (60 mM).

An alternate source of reduced nitrogen favoring embryogenesis can be an undefined substance like coconut milk,[328,379] casein hydrolysate,[9] or a mixture of amino acids.[161] Individual amino acid, L-glutamine or L-alanine, promotes embryo formation.[407] Glutamine is specifically important; it readily promotes embryo formation.[156] Its effect is also evident in *Dioscorea*,[8] *Gossypium*,[258] and *Carica*,[191] even when included along with ammonium. When α-alanine is added to nutrient medium, it is quickly transformed to glutamic acid by alanineamino transferase and utilized as a nitrogen source.[158]

Very little is known about factors which commit the cells to differentiate and form embryos.[300] In *Daucus carota*, inclusion of amino acids, serine, and proline along with auxin in callus multiplication medium, specifically promoted embryogenesis when callus was transfered to embryo-forming, hormone-free medium.[285] Since proline and serine are major constituents of cell wall glycoprotein, they are ascribed to have a morphoregulatory function. Also, in *Medicago sativa*[305] embryo formation is increased five- to tenfold by alanine and proline. However, utilization of nitrogen for synthesis of protein from alanine, proline, glutamine, and glycine is not qualitatively different, even though the latter two amino acids do not increase embryo formation. Therefore, further work is required to resolve the specific role of amino acids in embryo formation. Also of interest in this context is the promotion of embryo formation in *Glycine max* cv. Williams 82 by thiamine hydrochloride and nicotinic acid.[21] The addition of 5.0 μM thiamine HCl increased embryogenesis from 33 to 58%, and the addition of 30 μM nicotinic acid enhanced embryogenesis further to 76%.

Of the other ions, potassium is essential. Its effect is more pronounced when nitrogen of the medium is low.[280] Therefore, the usual practice is to prefer high salt medium of Murashige and Skoog[228] for embryogenesis than low salt medium of White.[409] It is believed that increasing reports of embryo formation in tissue cultures of diverse plants are due to use of high salt media as compared to low salt media employed earlier.

ii. Auxin

An increase in the number of embryogenic cells is possible during tissue proliferation on an auxin medium.[125,126] On an auxin-rich proliferation medium appear localized groups of meristematic cells which keep on multiplying with the growth of callus. These cells are known as embryogenic cells and tissues formed as embryogenic clumps (EC).

The auxin most commonly employed for tissue proliferation is 2,4-D for dicots as well as monocots. Effective concentration range is 0.5 to 25 μM, but rarely very high concentration is employed, 452 μM for *Phoenix dactylifera*.[282] Other auxins rarely employed are NAA and IBA for pumpkin,[146] 2-benzothiozole acetic acid (BOTA) for *Tylophora indica*,[268] parachlorophenoxy acetic acid (PCPA) for *Allium sativa*,[3] and picloram for *Trifolium pratense*.[62] It will be of interest to probe into the specific role of different auxins in embryogenesis. Picloram is of particular interest for embryogenesis of legumes. In many earlier investigations on pea tissues, embryo formation did not occur, but on employing a very low concentration of picloram (0.06 mg/ℓ) it was possible to have embryos from leaf tissue of pea.[141]

However, for embryogenesis to proceed, the tissue is to be transferred to auxin-free or low auxin medium. In carrot tissue,[97,122] auxins 2,4-D and IAA inhibited embryogenesis and in *Coffea arabica*,[316] embryos appeared only when the callus was transferred to auxin-free medium. Although auxins are inhibitory, antiauxins PCIB and 2,4,6-T did not promote embryogenesis in carrot[98] and were rather inhibitory.

From the above account one can conclude that to begin with, there is need of auxin for the proliferation of embryogenic cells and, in turn, the process of embryogenesis is inhibited by auxin. This apparent complex role of auxin[99] was resolved when cell clusters of different dimensions were taken for embryogenesis. Cell clusters of size smaller than 31 μm rarely developed directly into embryos on an auxin-free medium. A supply of 2,4-D prior to induction of embryos enhanced embryo formation from these clusters when they were transferred to auxin-free medium. When cell clusters bigger than 31 to 47 μm were employed they could form embryos without any growth regulator. The role of hormones in regulation of embryo formation was further resolved when cultures were started from single cells.[243] Single cells from carrot cell suspension were filtered through sieving and then further fractionated employing density gradient centrifugation in percoll solution; 85 to 90% of these single cells resulted in the formation of embryos when cultured in a medium containing 2,4-D (5 × 10^{-8} M), zeatin (10^{-6} M), and mannitol (0.2 M) for 7 days followed by a transfer to a medium containing zeatin (10^{-7} M), but no auxin. This indicates that there are at least two phases in the differentiation of embryos from single cells. The progression of first phase requires an auxin whereas that of second phase is inhibited by auxin.

Promotion of a callus by an auxin that is capable of embryo formation is more clearly evident in monocots, particularly cereals. Immature embryos of wheat on culture on an auxin medium produce two types of calli.[247] One, embryogenic, consists of small meristematic cells and forms embryos. The other, nonembryogenic, consists of long tubular cells which gives rise to a few embryos. Also, the high concentration of tryptophan increased the formation of embryogenic callus in some cultivars of rice[304] (Calrose 76, Pokkali, and IR 36), but not in others (Mashuri, BG, H4, and Giza). The difference between Japonica (Calrose 76 and Giza 159) and indica (Pokkali and IR 36) cultivars is not the causal factor for the difference in response to tryptophan.

A high level of endogenous auxin inhibits embryogenesis. This is best seen in a work on habituated tissue of *Citrus sinensis*. Originally, *Citrus*[170] tissue raised from nucellus required auxin as well as cytokinin for cell proliferation and embryo formation. However, on repeated subculture the tissue gradually lost its embryogenic potential and from it, some cell lines could be isolated which were hormone autotrophic. In these cell lines promotion of embryogenesis was possible by antiauxins, 2-hydroxy-5-nitrobenzyl bromide, and 7-azaindole.[167-169]

Other growth regulators, cytokinis and gibberellin, are ineffective in promotion of embryogenesis. Cytokinin[124] is rather described to favor proliferation of nonembryogenic cells in a callus. Also, a determination of endogenous levels of cytokinin nucleotides during embryo formation in *Pimpinella* (anise) cell cultures indicated that cytokinins have a major role in cell division and not in cell differentiation.[83] However, zeatin alone and not kinetin or benzylaminopurine promoted embryo formation in *D. carota*.[99] Also, callus culture from leaves of *C. arabica*[316] required high cytokinin to auxin ratio for proliferation and manifestation of embryogenic potential on auxin-free medium. Another interesting exception is promotion of embryogenesis in ovular callus of *Citrus*.[171] More recently, in *Coffea arabica*[418] it has been shown that leaf explants produce friable white callus on medium containing benzyladenine as a sole source of growth regulator and this tissue forms embryos.

As for gibberellins, they inhibit embryo formation in carrot and anise and this inhibition is correlated with a high level of endogenous gibberellins and a reduced rate of gibberellin metabolism.[241] As an exception, gibberellic acid promoted embryo formation in one cell line of *Theobroma cacoa* at 0.5 to 10 mg/ℓ.[179] However, it was ineffective in another cell line. It is not clear how gibberellic acid promotes embryogenesis, because this was also possible with AMO 1618, an inhibitor of gibberellin biosynthesis, at low concentrations. Other inhibitors, diaminozide and CCC, repressed embryogenesis.

ABA does not promote embryo formation in carrot.[157] As an exception, it promoted embryo formation from habituated tissue of *Citrus*.[171]

b. Tissue Origin

Carrot and *Ranunculus* are exceptional in showing high propensity of embryo formation from almost all parts of the plant. Of these, carrot has been subjected to detailed investigations. In this plant, embryo formation has been reported from tissues obtained from tap root,[327] embryo,[326] peduncle,[125] petiole,[122] young root,[315] and hypocotyl.[98]

Contrary to carrot, tissue specificity and embryo formation can be seen in *Citrus*. In different species of this genus from the nucellus in vivo arise adventive embryos which develop into plantlets and serve for propagation of elite plants. Also in vitro, nucellus is a good source of embryo-forming tissue. Nucellar tissue from ovules of *Citrus microcarpa*[266] and *C. reticulata*[290] readily formed embryos and only tissue from the micropylar end was specific to form embryos. Further, the tissue from nucellus of postfertilized ovules would differentiate embryos in culture. However, specificity of postfertilized ovules was not seen in embryo formation from nucellar tissue of citrus cultivars, Washington naval orange,[48] and Shamouti orange.[169] The work described above is based on species which show nucellar polyembryony in nature, but nucellus tissue from *Citrus* species, which does not show polyembryony in nature, is also capable of differentiating embryos in culture.[265] Other examples of tissue specificity are peduncle of *Carica stipulata*,[191] immature ovary of *Hemerocallis*,[181] and ovules of *Vitis vinifera*[318] and *Carica papaya*.[192]

Tissue specificity and embryo formation is characteristically apparent in members of graminae. Preferentially, tissue from young inflorescences or immature embryos differentiate to produce embryos (Figure 6). This is true for grasses *Bromus*,[103] *Lolium*,[70] *Echinochloa* spp.,[340,380,395] *Dendrocalamus*,[267] and *Cyanodon*,[5] millets *Pennisetum americanum*,[388] *P. purpureum*,[129,397] *Sorghum*,[33] *Zea mays*,[77,197,261,269] and cereals *Triticum aestivum*[185,247] *Oryza*

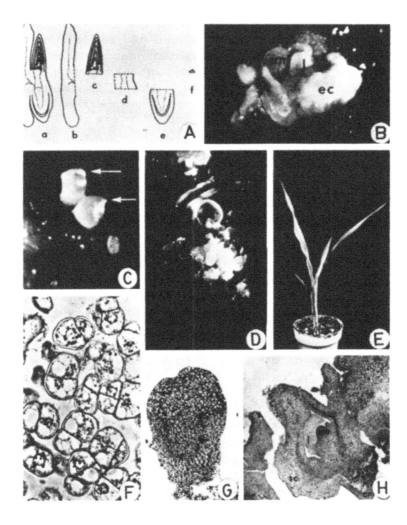

FIGURE 6.   Embryo development and shoot formation on culture of different parts of mature embryos of *Pennisetum americanum*. (A) Explants for culture; a, complete embryo; b, scutellum; c, shoot-axis; d, mesocotyl; e, root-apex with coleorhiza; f, shoot-apex with two youngest leaf primordia. (B) Embryo on culture on auxin medium showing initiation of embryogenic callus (ec) at the base of leaf primordium. (C) Embryos differentiated from semifriable callus. (D) Shoot developed from a somatic embryo. (E) Plantlet regenerated from a somatic embryo. (F) Embryogenic cells from a compact callus. (G) Globular embryo. (H) Well-developed embryo. (From Botti, C. and Vasil, I. K., Z. *Pflanzenphysiol.*, 111, 319, 1983. With permission.)

sativa,[55,131,190] *Secale cereale*,[198] and *Hordeum vulgare*.[359] To an extent, cells from immature basal parts of leaf also form a tissue which is capable of embryogenesis in Sorghum,[403] rice,[25,404] napier grass,[129,397] and wheat.[405] In wheat the response was restricted to certain genotypes. Reports of recent origin indicate that regular plant formation is also possible from tissues originating from other parts: apical meristem in barley,[399] root tip in rice,[1] and glume in maize.[335]

Similar to graminae are seed legumes, which are relatively refractory to regeneration. A low to infrequent embryogenic response is reported in some seed legumes including *Glycine max*. Plant regeneration has been possible in *G. canescens*,[113,236,412] a wild relative of soybean, and *Vigna aconitifolia*,[24,109] a drought-resistant primitive seed legume. However, of late, even in soybean plant regeneration has been possible from 54 cultivars from tissue derived either from immature embryos[21] or cotyledonary node,[164] indicating thereby that tissue specifity is a critical determinant in regeneration.

Not only tissue specificity, but cell specificity is also seen. In *Triticum*, only tissue derived from cells of scutellum of an immature embryo is capable of differentiating into embryos. Therefore, the position of an embryo during culture is an essential factor for success. The embryonic shoot-root axis should be in contact with the medium and scutellum on top of it proliferates to form a callus which is capable of regeneration. Further, not all cells of nucellus are capable of producing a tissue which differentiates into embryos. Only compact callus originating from epiblast cells of nucellus is morphogenic and friable tissue arising from other cells is nonmorphogenic.[248] The origin of embryos or embryogenic tissue from scutellum needs to be looked into in greater detail. On culture of immature embryos of wheat[202] cultivars Froid centurk and Helge embryos could be observed as early as 6 days, and their origin was traceable to three basic tissues of scutellum: (1) epithelial cells divided to form a thallus type of structure which acquired the bipolar symmetry of an embryo, (2) in actively dividing ground tissue, localized divisions in some cells resulted into embryos, and (3) divisions in some of the procambial cells of nucellus resulted in formation of roots. Also in *Zea mays*[387] the embryos arise from epidermal and subepidermal cells of scutellum.

3. Embryogenesis — An Overview

In addition to the differentiation of embryos from callus tissue, embryo formation is also of frequent occurrence on hypocotyl of plantlets, formed in vitro from embryos, of *Ranunculus sceleratus*,[176] *Daucus carota*,[161] *Atropa belladonna*,[272] and *Medicago sativa*.[199,200] As many as 5 to 50 embryos were seen on one plantlet in *R. sceleratus* (Figure 7). The origin of stem embryos in this instance is directly from individual cells.[178] The appearance of embryos on stems of in vitro formed plantlets is possible on mineral medium lacking hormones. These are likely examples of predetermined embryogenic cells. To be included in this category are the instances of embryo formation on cotyledons of *Ilex aquifolium*[133,134] and *Juglans regia*.[378]

Thus, embryogenesis can follow two routes, direct embryogenesis, where embryo formation initiates from cells without callus formation, or indirect embryogenesis, in which cell proliferation occurs prior to initiation of embryos. This pathway is followed by callus cells. It is very likely that in indirect embryogenesis cells are to be induced towards embryo formation, whereas direct embryogenesis is possible from cells which are either determined or can be directed towards embryo formation. This is consistent with the direct induction of embryogenesis, without callus formation, from mesophyll cells of orchard grass[64] in response to 3,6-dichloro-*o*-anisic acid, dicamba. Also of interest in this context are recent observations on immature embryos of *Trifolium*. When immature sexual embryos of *T. repens*[204] were cultured on benzylaminopurine medium, embryos differentiated directly from hypocotyl axis. The first sign of induction was the shift from regular anticlinal division to irregular periclinal and oblique division. The effect of BAP is not entry of cells into mitosis per se, but an alteration[413] of cell polarity and division plane leading to formation of embryos.

C. Formation of Flowers

The formation of flower-buds under cultural conditions is of infrequent occurrence compared to formation of shoot-buds. Flower formation was first observed on explants from inflorescence regions of photoperiodic insensitive cultivar of *Nicotiana tabacum*.[4] There was a close correlation between physiological state of explants and the response. Stem segments from basal parts of the plant formed vegetative buds and segments from inflorescence region produced floral buds. These findings are confirmed in other plants which are either cold-requiring or photoperiodic. In *Chicorium intybus*[238] and *Lunaria annua*[255] flower formation was possible by vernalization of explants whereas in *Plumbago indica*,[237] flower formation

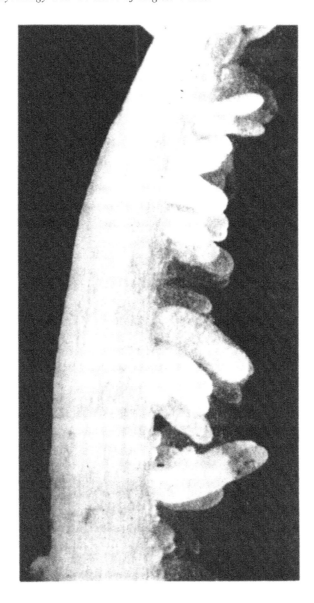

FIGURE 7.   Embryos on hypocotyl of plantlet of *Ranunculus sceleratus* formed in vitro. Courtesy of the Late Dr. R. N. Konar)

on stem segments was possible in short days. In addition to these basic requirements, high
levels of nitrogen, cytokinin, adenine, and orotic acid were required. Auxin and gibberellin
were inhibitory. Auxin affects the floral gradient in *N. tabacum.* On increasing the auxin
concentration there is complete abolition of flower-bud formation on basal segments.

Perilla frutescens[344] is another system where a gradient of floral bud-forming capacity
can be seen along the stem. Interestingly, in this system occasional floral bud formation
was also possible on leaf discs and it depended on the position of leaf on the plant.

Yet another system is *Torenia,*[345] where flower formation is possible on inflorescence
segments. In this system, elimination of ammonium nitrate from culture medium promoted
flowering, and dilution of other salts and an increase in carbohydrates were also helpful. In
contrast to flower formation in *Nicotiana, Chicorium, Lunaria,* and *Plumbago* described
above, flower formation on stem segments of *Torenia*[346] was stimulated by auxin. The effect

was pronounced when inflorescence explants were taken from plants during the flower abscission stage. However, when explants were taken from younger plants, the effect was insignificant. Cytokinin, zeatin, at 1 mg/ℓ inhibited floral bud formation, but it promoted flowering on basal internodes or explants from plants in the late stage of flowering. An in-depth study of flower formation in *Torenia*[347] stem segments was possible due to division of response into three different phases. During the first phase, 2 weeks, adventitious buds are initiated. In the second phase, 3 to 4 weeks, the floral-buds differentiate, and in the third phase, 5 to 12 weeks, floral buds develop. In this system, floral bud formation is possible in the absence of ammonium nitrate and growth regulators. Ammonium nitrate and zeatin suppressed floral bud development when given during the third phase. However, zeatin slightly promoted flowering if given during the first phase. Auxin, IAA, was stimulatory to floral bud formation only when given during the first phase. Interestingly, in *Torenia*, application of ABA (100 mg/mℓ) stimulated flower formation on stem segments from vegetative plants which had no predisposition to fertility.[350] A higher rate of flower formation was obtained when endogenous content of ABA (resulting from native and external supply) was between 16 to 20 μg/g fresh weight. This clearly indicates that the endogenous level of ABA is the critical factor controlling flowering in this plant.

More interesting are reports of direct formation of flowers on cotyledon of *Torenia*[376] *Carthamus tinctorius*,[353] and *Arachis hypogea*.[233] In *C. tinctorius*, from the inner surface of cotyledon developed capitula in which complete blooming of florets was seen within 55 to 90 days. In *A. hypogea*, when cotyledons (Figure 8) with embryo axis were cultured on cytokinin medium, shoots were produced in the axils of which two to seven flower buds could be seen. The number of flower buds increased with increasing concentration of cytokinin, the optimal level being 3 mg/ℓ of kinetin or 4 mg/ℓ of BAP. When cotyledons alone were cultured, 8 to 28 flower buds developed directly on them, at 0.5 mg/ℓ of BAP, without any vegetative growth. Excised embryo axis on culture on the same medium gave a plant without any flower bud. Auxins and gibberellin had no effect on flower bud production. A low percentage of these flowers formed gynophores and, ultimately, pods.

Regeneration of floral buds is also possible from root segments of *Cichorium intybus*.[30,31] Root segments formed shoot buds on a filter paper support in a liquid medium, and when supported on agar medium, formed floral buds. *Cichorium intybus* is the only known species able to form flowers on root explants after cold and long-day treatments. There is evidence for involvement of phytochrome (Figure 9). Red light promoted flowering while FR and a combination of red and FR had no effect.[11] In short-day conditions, floral response could be obtained by interrupting dark period with brief irradiations of red light; these were counteracted by FR irradiation.

D. Multiprogrammable Morphogenesis

From the foregoing account it becomes clear that from an unorganized tissue or a callus only a limited potential of differentiation is possible. The regenerants are either embryos, shoots, or roots. Formation of flowers from callus culture of *Phlox*[175] and *Crepis*[279] are rare instances. So far, induction of floral buds has not been reported from a callus tissue.

By contrast, in a thin-layer system, a multiprogrammable potentiality of regeneration is possible. A thin-layer system is comprised of an epidermis and a few cell layers of cortex. Thin-layer explants from *Begonia, Nautilocalyx, Saintpaulia, Brassica, Catharanthus, Solanum, Torenia, Nicotiana* and *Cichorium* formed roots, shoot buds, and callus. However, in *Nicotiana, Cichorium,* and *Torenia*, flower buds were also possible. A new morphogenic expression, formation of unicellular hair, was also possible in *Begonia*. Therefore, multiprogrammable morphogenesis is possible employing the thin-layer system.[376]

Studies on the thin-layer system began with *N. tabacum* cv. Wisconsin 38[374,375] and it has been most intensively investigated. Basically, the morphogenic expression in cultures

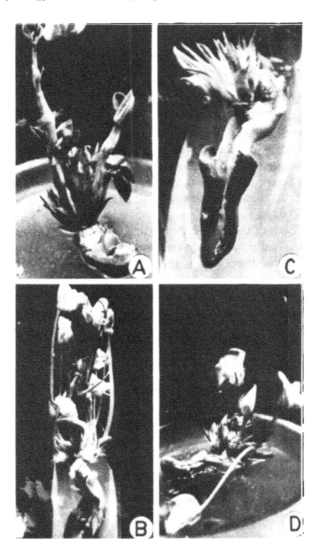

FIGURE 8.   Flower formation from cotyledons of *Arachis hypogea*.
(A) Cotyledon with an embryo axis on kinetin medium showing growth
of shoot axis and formation of flowers. (B) A similar culture, showing
blooming of flowers. (C) Cotyledon, without an embryo axis, showing
direct differentiation of flower buds on kinetin medium. (D) Similar
culture, showing blooming. (From Narasimhulu, S. B. and Reddy,
G. M., *Theor. Appl. Genet.*, 69, 87, 1984. With permission.)

of thin layers depends upon the region of the plant from which thin layers are excised. The
peels derived from floral branches normally produce floral buds (Figure 10) and those from
basal parts produce vegetative buds, whereas peels from the intermediate region produce
both types of buds in different proportions.

The morphogenic expression of peel segments taken from the inflorescence region can
be modified, at will, to produce either floral buds or vegetative buds, or roots or none —
only callus or no division at all. The peel segment is not only a multiprogrammable system,
but in it one can have morphogenic expression in a relatively short period of 8 to 12 days.
In a thin-layer system it is possible to observe changes taking place in individual cells leading
to different types of morphogenic patterns. Also, in this system the parent tissue is reduced

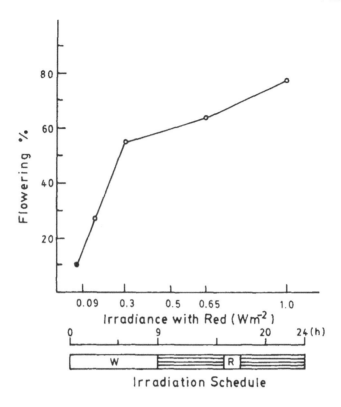

FIGURE 9. In vitro flower formation of *Chicorium*, in response to red light. Red light was given for 15 hr in addition to 9 hr white light. (From Badila, P., Lauzac, M., and Paulet, P., *Physiol. Plant.*, 65, 305, 1985. With permission.)

to a minimum, consequently the carryover of endogenous factors is also reduced to a minimum and it is possible to correlate a specific factor with a specific response. The thin-layer system can provide a deeper insight into the understanding of morphogenic processes because unlike callus system, the "target" cells are not scattered among the heterogenous mass of cells. However, the small size of thin-layer can be a disadvantage for biochemical studies.

1. Factors Affecting Morphogenesis in Peel Segments

Of the factors affecting flower formation in thin-layer explants, the most significant is the flowering stage of parent plant when thin layers are excised from the inflorescence region. Optimal response, 100% of the peels forming flowers, was possible when peels were taken from plants bearing green fruits. When donor plants had reached mature fruit stage, the response declined to 85%. There was no response if the parent plant had only flowers and not fruits.

Under identical conditions favoring floral bud formation, the epidermal peels could be made to form 100% vegetative buds if they were floated on a liquid medium.[66] Controls on agar formed floral buds.

The peel segments from the inflorescence region of tobacco plant produced floral buds at a low and an equimolar level of cytokinin and auxin along with 30 g/ℓ of glucose. In the presence of hormones, but in the absence of glucose, floral buds did not appear and only after 18 days did a few shoot buds appear. Carbohydrate was, however, not required during the first 4 days, but lack of glucose from 6 to 10 days resulted in abnormal meristems.

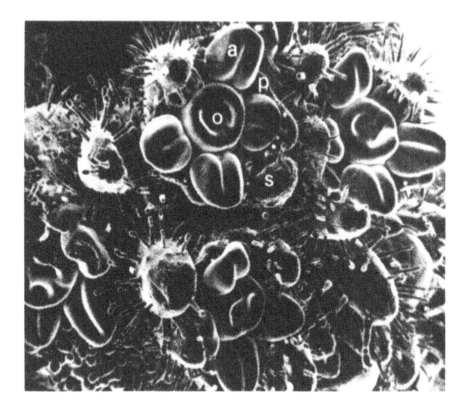

FIGURE 10.   Peel segment of *Nicotiana tabacum* showing direct differentiation of flower buds, without an intervening callus phase. a, anther; p, petal; s, sepal. (After Tran Thanh Van. K., *Int. Rev. Cytol.*, 11A, 175, 1980. With permission.)

Light is also an absolute requirement for differentiation of floral buds and the critical phase during which light is required is from day 7 to day 10. On transfer of culture to dark on 7th day the floral buds could be converted to vegetative buds. This critical phase of light requirement coincides with the requirement for glucose. Under continuous or sequential deprivation of sugar, either no organogenesis occurs or abnormal structures or vegetative buds are formed. Therefore light per se is not sufficient for floral bud induction.[67,68] Thymidine analog, 5-bromodeoxyuridine (Budr) specifically inhibited flower formation when applied for 24 hr at the 8-day stage. An increase in the level of cytokinin without a change of auxin level, in the presence or absence of glucose, resulted in the formation of vegetative buds. By further alterations in the level of hormone it was possible to have either roots or unorganized callus. Therefore, one can evoke the determination, reversibility, and trans-determination and draw a parallel[377] with differentiation in *Drosophila*. The possibility of a sequential supply of glucose or light and inhibition of flower bud formation by Budr at a specific stage speaks of a determinative phase in this morphogenic process. This is identical to stage-specific inhibition of organogenesis, phenocritical times, described in Section IV. A. 2. a.

About the hormonal requirement for floral bud differentiation, cytokinin can be described to be absolute. Auxin acts as modifier in a complex way. Auxin, NAA, at 1 μM in early stages decreased the number of initiated floral buds and delayed their emergence. However, at a later stage, at the same concentration, it promoted the outgrowth of floral buds. Optimal results[81,82] were possible when peels were cultured on a low level of auxin for 3 to 5 days and then transferred to elevated level of auxin. Organogenesis of flower buds is completed within 10 to 14 days. The hormones NAA and BAP in different concentration affect the formation and distribution of flower buds, bud morphology, and callus formation. In particular, BAP influences bud formation and bud morphology while NAA affects callus formation. The role of auxin in flower formation is also minimized by the observation that there is a gradient in flower-forming ability of explants, but levels of IAA in these tissues did not show this gradient.[242]

V. REGENERATION FROM INTACT TISSUE

Control of morphogenesis can be studied employing plant segments which regenerate directly to form shoots or roots, without an intervening phase of dedifferentiation or callus formation. In nature, segments from root, stem, and leaf of several taxa[73] are able to differentiate shoots and form new individuals. In some instances a mere injury or isolation is sufficient to trigger shoot bud or root formation. In many of the perennial legumes regeneration of shoots from a root stock is an annual feature. Rarely, on a plant with high regeneration potential, such as *Begonia*, it is possible to induce *de novo* shoot bud formation on leaves, while they are attached to plants, by the application of a cytokinin, 6-7-7-dimethylallylaminopurine.[56] Regeneration of plants from intact tissues has been possible employing either shoot apices or explants from shoot, root, leaf, or endosperm.

A. Regeneration from Shoot Apex

Regeneration of plants on culture of shoot apices has been possible in a large number of plants,[224] and this is considered to be an alternative method for micropropagation. In particular, a recourse to this technique has been found to be essential for raising virus-free plants and micropropagation of orchids. However, before taking the specific examples it is in order to know the requirements for regeneration of a meristem.

1. Requirements for Regeneration of a Meristem

Contrary to earlier reports that shoot apices of Angiosperms lack the potential for autonomous growth on culture, it has been unequivocally demonstrated that shoot apices of several species, comprising an apical dome only, free of subjacent stem tissue and leaf primordia, are capable of regenerating into entire plants on a simple mineral-sucrose medium.[312] This has been confirmed in subsequent investigations.[14,283] Further, it has also been resolved that there are physiological differences in the region of the shoot apex. On a medium devoid of growth regulators explants prepared from different regions of meristem gave variable results. The central cells produced fewer mature leaves and fewer embryos than the cells from flanks.[14]

Some genera are, however, exceptions to the generalization that regeneration is possible on simple mineral-sucrose medium. For instance, in *Coleus blumie*[311] the regeneration was possible only in the presence of IAA or in the presence of some leaf primordia. Also, in *Dianthus caryophyllus*[299] in the absence of hormone, IAA, or kinetin, two pairs of primordial and a pair of expanding leaves were required for plant regeneration. Similarly, apical meristems of *Pharbitis nil*[17] could regenerate into plants in the presence of auxin. In *Coleus blumei*[313] it is further resolved that in presence of IAA, two pairs of primordial leaves were essential for plant regeneration whereas with IAA and kinetin, one pair of primordial leaf was enough. It appears that developing primordial leaves supply to the apical meristem phytohormones required for growth.

2. Micropropagation on Culture of Shoot Apex

Meristematic tissues in general are considered to have a relatively higher regeneration potential. Compared to forage legumes, seed legumes are relatively refractory to regeneration. Raising of tissue cultures in pea has been possible from roots[373] and shoot apices,[104] but plant regeneration was possible only in tissue derived from shoot apices.

Shoot tip culture has also been found to be an efficient method for clonal multiplication of grape plants.[18] When a shoot tip measuring about 1 mm was cut into 20 smaller fragments, each fragment formed multiple shoots on mineral medium supplemented with sucrose and cytokinin. The shoots formed in vitro could be employed for subsequent culture. In brief, starting with one shoot apex, with an interval of 10 to 15 days between different experiments, within 4 months it was possible to have 8000 plants. Similarly, from one shoot apex of *Rosmarinus officinalis*[53] it was possible to have 5000 plants per year. About 10 to 15 shoot buds differentiated from one shoot apex.

Shoot apices also show cytological stability during differentiation in vitro. This is in contrast to cell cultures and is possibly due to the high division capacity of shoot apex cells. Plants regenerated from shoot apices of *Asparagus*[227] *and Gerbera*[226] were diploid.

3. Culture of Shoot Apex and Propagation of Virus-Free Plant

Virus-free plants are also possible on the culture of shoot apices. This is because the titer of virus increases with increasing distance from meristem tip.[132,160] The apical meristem is either free of virus or has a vey low titer.[260,396] A low titer or absence of virus in shoot apices is due to (1) lack of vascular tissue in shoot apex, which prevents the movement of virus and (2) high auxin level in shoot apex, which is inhibitory to virus multiplication.

Employing shoot tip culture, it has been possible to raise virus-free plants in many ornamentals such as *Dahlia*,[132,219,260] *Cymbidium*,[217] *Fressia*,[34,35] *Iris*,[20] *Dianthus*,[330] *Gladiolus*,[302] *Hippaestrum*,[244] *Narcissus*,[249,331] *Nasturtium*,[394] and *Lavendula*.[393] This has also been possible in crop plants such as apple,[49] sugar cane,[184,220] cauliflower,[249] taro,[128] strawberry,[220] soybean,[408] sweet potato,[220] potato,[220] cassava,[154] and ginger.[396]

In order to obtain a virus-free plant it is necessary that culture is initiated with the submillimeter portion from extreme tip.

4. Culture of Shoot Apex and Propagation of Orchids

Many of the cultivated orchids are complex hybrids and in the past, their multiplication by seeds involved a high degree of risk due to ensuing heterogenity in the progeny. Consequently, orchid lovers had to pay a prohibitive price for a desired plant. However, thanks to a French botanist, Morel,[217] it is now possible to have clonal propagation of orchids, at an astronomical rate, on culture of a shoot apex.

When the shoot apex from a plant of *Cymbidium*[217,218] was cultured on Kundson's mineral agar medium, containing 2% sucrose as the only organic source, it formed protocorm-like spherical body with rhizoids at the base. It was similar to the protocorm produced by germinating orchid embryo. From this protocorm developed two to five secondary protocorms in a matter of 1 to 2 months and each of the protocorms developed into a plant. When one of the protocorms was isolated and cut into segments each segment developed into three to six protocorms. Thus, starting from a single shoot apex it was possible to have indefinite multiplication of plants on a simple mineral-sucrose medium (Figure 11).

Employing this technique, now orchids, sympodial as well as monopodial, are routinely multiplied at industrial scale in many countries. It has been possible to multiply almost all economically important orchids tried so far, except *Paphiopedilum*. In this orchid it has not been possible to induce the formation of protocorm.

B. Regeneration from Stem Segment

In nature, *Linum usitatissimum* is a good example of shoot bud formation on intact hypocotyl.[69] On culture, a 15-mm-long hypocotyl segment of *Linum*[229] may give rise to 100 to 170 potential shoot buds. In vitro regeneration of shoot buds has also been possible from stem segments of *Brassica*,[159] *Chrysanthemum*,[284] and *Rudbeckia*.

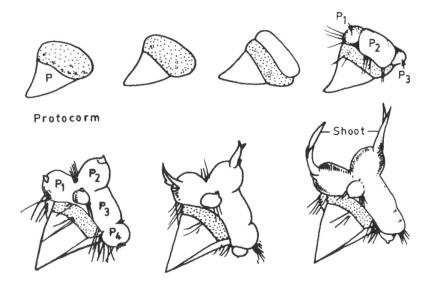

Protocorm

Shoot

FIGURE 11. Regeneration of several protocorms and their shoot formation from a single protocorm fragment. (Redrawn after Morel, G., *Phytomorphology*, 22, 265, 1975. With permission.)

The physiological role of auxin and cytokinin in induction of regeneration has been studied in detail in *Torenia*[348] stem segments. Cytokinin alone can induce shoot bud formation in *Torenia*. The number of meristematic zones and buds arising from it increased with increasing concentration of cytokinin. The response by zeatin was suppressed by simultaneous application of auxin or anticytokinin. Antiauxin at a low concentration increased the response; this was possible with or without cytokinin. Root formation was possible when explants were treated with auxin, NAA or IAA, at a concentration higher than 0.05 μM. Antiauxin and cytokinin significantly reduced the number of roots, anticytokinin along with auxin, promoted root formation. The induction of adventitious buds by cytokinin on stem segments of *Torenia* is possibly mediated by an increase in the intracellular level of calcium.[349] Meristematic divisions on stem segments, which lead to shoot bud formation, was possible in the absence of cytokinin, but on application of calcium ionophore, A23187. In the absence of calcium and on medium containing EGTA, cytokinin failed to induce shoot bud formation. Similar inhibition was also possible by lanthanum, a calcium antagonist, verapamil, a calcium channel inhibitor, and trifluoperazine and chlorpromazine, calmodulin inhibitors.

The shoot-forming capacity of stem segments of *Populus*[76] is dependent on the size of explants. Segments 6 mm in length formed adventitious buds on basal medium containing only sucrose, whereas segments 2 mm in length were nonmorphogenic. However, the addition of 0.1 mg/ℓ of zeatin was helpful in stimulating shoot bud formation.

C. Regeneration from Leaf Segment

Propagation by leaf segments is a standard practice in many plants. There are about 350 species in which leaves are capable of producing adventitious buds.[37] More common are *Begonia, Crassula, Pepromia, Pinguicula,* and *Saintpaulia*.

Leaves of *Begonia* are known for their regenerative capacity to produce shoot buds from cut ends. Even leaves attached on plants produced shoot buds on application of cytokinin.[56] On culture of leaf segments on cytokinin medium the regeneration was so profuse that entire surface was covered with shoot buds.[281] Surprisingly, even ABA stimulated shoot bud formation on leaf segments of *Begonia*.[130] Large-scale propagation of *Begonia* × *hiemalis* plantlets was possible on culture of leaf segments in bubble column-type liquid shake culture.[341]

Regeneration from leaf explants is described to be a function of leaf age. Explants from younger leaves of *Echeveria*[264] regenerated roots and those from older leaves only shoot buds.

Leaf discs of tomato also form shoot buds.[23] In a combination of auxin and cytokinin it was possible to have either callus, shoots, roots, or shoots and roots. The tomato leaf disc system has been employed for studying the role of ethylene in organogenesis.[61] At an increased IAA level there was an increased rooting response and it correlated with increased ethylene production, but the external supply of ethylene or ethepon did not stimulate rooting from leaf discs. Also, absorption of ethylene, evolved by leaf discs by mercuric perchlorate and pretreatment of leaf discs by AgNO$_3$, an inhibitor of ethylene action, significantly promoted IAA-induced rooting response. Therefore, ethylene as such does not appear to be a rooting hormone in tomato.

Variable shoot-forming response of different species of a genus in respect of different concentration of auxin and cytokinin is on record in *Brassica*.[78] For an optimal response *B. oleracea* required 10 mg/ℓ BAP and 1 mg/ℓ NAA; it was 10 mg/ℓ BAP and 10 mg/ℓ NAA for *B. napus*, and *B. campestris* required 1 mg/ℓ of BAP and 10 mg/ℓ of NAA.

A systematic study of effects of auxin, cytokinin, and GA on regenerative capacity of leaf explants of *Digitalis purpurea*[288] showed that IAA stimulated root formation and it was inhibited by BAP. Higher concentrations of GA$_3$ were able to stimulate or inhibit root formation depending on how low or high IAA level was present in the medium. For shoot bud formation BAP and IAA were indispensable. Although GA$_3$ was not required for the response, it stimulated shoot bud formation at the low level, but was inhibitory at a higher level. Essentially similar results were obtained from leaf explants of *D. obscura*.[251]

More interesting is the recent work on regeneration of leaf discs of *Perilla*, *Hoya*, and *Solanum melongena*. Explants from leaves of *Perilla*[344] formed embryos when the material was first grown on 2,4-D medium and transferred to auxin-free medium. Cytokinin, *N*-phenyl-*N'*-(4 pyridyl) urea and its derivatives induced shoot buds. Auxins, NAA, and NOA at a low concentration induced shoot bud formation. At a high concentration NAA induced compact callus at cut ends, and 2,4-D induced a friable callus. Similarly, from leaf discs of *Hoya*[206] embryo formation was possible on medium containing 2,4-D and kinetin and not on media containing IAA and kinetin or NAA and kinetin. However, leaf explants of *S. melongena*[108] formed callus and embryos on NAA, whereas other auxins induced only callusing. Maximum embryo formation occurred at 10 mg/ℓ of NAA. Cytokinin inhibited NAA-induced embryo formation, but in combination with NAA, promoted callus growth. The frequency of embryo formation was under the influence of nitrogen content of medium; both NO$_3^-$ and NH$_4^+$ were essential for embryo formation in a ratio of 2:1. Sucrose level was also important; the optimal level was 0.06 *M*, and both elevated and reduced levels were inhibitory to embryo formation. Only in the presence of cytokinin was shoot bud formation possible.

For the propagation of orchids, culture of shoot apex is the standard practice. However, this technique requires the sacrifice of a bud or even a whole plant in case of monopodial genus. Therefore, as an alternative, regeneration is sought from leaf explants. Leaf explants of *Cymbidium* also produce protocorm-like bodies. Similarly protocorm-like bodies were also obtained from leaf segments of *Rhyncostylis*,[390] a monopodial orchid. Regeneration was possible for immature leaves only; explants from mature leaves necrosed.

D. Regeneration from Root Segment

Since the original work on the establishment of root culture of tomato,[409] roots of many plants are shown to be capable of independent growth and regeneration to form secondary roots in culture.[47] Regeneration of shoot buds from isolated roots on culture has, however, been observed in relatively few cases and mainly in those species which form shoot buds on roots in vivo.[253] The shoot buds arise *de novo* from pericycle, as in *Convolvulus*.[372]

Roots of many solanaceous plants are capable of regenerating shoot buds. Root cultures of *Atropa belladonna*[357] formed shoot buds on aging of cultures. Numerous embryoids were possible on root culture of *Solanum khasianum*[54] in liquid medium containing BAP, while on the same medium with agar shoot buds differentiated. Roots of *Nicotiana exigua, N. debneyi, Solanum melongena,*[420] and *S. tuberosum*[84] formed shoot buds on mineral medium without any growth regulator. The potentiality was maintained through several subcultures of root segments. It is interesting to know that root of *N. tabacum*, a plant of high regenerative potential, does not form shoot buds in culture. Regeneration of shoot buds has also been reported on root segments of *Brassica*[186] on medium containing auxin and cytokinin; kinetin and picloram gave the highest frequency of regeneration.

An interesting case of differential regeneration capacity of roots regenerated from male and female plants was seen in *Actinidia chinensis*.[38] Roots of female origin were relatively less responsive to shoot bud formation whereas roots of male plant origin showed high capacity of regeneration to form shoot buds.

Multiplication of orchids is also possible on culture of root. Regeneration of protocorm-like bodies was seen on culture of root tips of *Catasetum*,[163] a brazilian orchid.

E. Regeneration from Endosperm

Endosperm is a specialized tissue — meant for nourishment of a developing embryo. Being triploid, it is also unique and is especially advantageous for raising triploid plants otherwise obtained by crossing of a tetraploid with a diploid. Triploids are desirable for raising seedless varieties. Many of our present day fruit crops — apple, banana, mulberry, and watermelon and other crops, such as sugarbeet and tea, are triploids obtained by breeding. In order to avoid crossing and rule out the uncertainty of success, a simple alternative is to employ endosperm for regeneration of triploid plants. Besides seedless fruits, triploids can have other agronomic qualities. Triploids of *Populus*[414] have better pulp and wood qualities and can be multiplied vegetatively.

Shoot formation from endosperm tissue was possible, for the first time, in a parasitic plant *Exocarpus cupressiformis*.[148] In cultures of whole seeds on medium containing IAA, kinetin, and casein hydrolysate (CH) about 10% of the explants formed shoot buds from endosperm. Since then, organogenesis and embryo formation from endosperm have been demonstrated in a number of parasitic plants and a few autotrophic plants. Interestingly, in parasitic species there is a direct differentiation of shoot buds from endosperm, and in autotrophic species the differentiation is preceded by callusing of endosperm.

Cytokinin is an essential requirement for shoot bud formation of *Scurrula pulverulenta*[26] and *Taxillus vestitus*,[149,230] and auxin, IAA, counteracts the effect. A mere soaking of endosperm pieces of *T. vestitus* in 0.025% kinetin for 24 hr resulted in formation of buds on mineral medium alone. In *Leptomeria acidia* and *Dendrophthoe falcata*,[230] kinetin-induced shoot bud formation occurred in the presence of a low level of auxin, IAA or IBA. An increase in the level of cytokinin resulted in an increased response, but a corresponding increase in auxin was of no help.

Of these genera, *Taxillus vestitus* shows a high propensity of regeneration from endosperm. The presence of an embryo does not favor regeneration. An increased response was more possible from endosperm halves than from an intact endosperm with embryo. It remains to be ascertained whether it is due to injury, because shoot buds invariably occurred along the injured region. When the split half of endosperm was planted with its cut surface in contact with the medium containing cytokinin (5 mg/ℓ), in 100% of the cultures 12 to 18 buds appeared per culture whereas in a reverse position, cut surface away from medium, only one to three buds appeared in 30% cultures.

In autotrophic species callus formation precedes differentiation. Tissues from endosperm of *Croton*,[27] *Jatropha*,[319] *Putranjiva*,[320] and *Annona*[232] remain undifferentiated on auxin-cytokinin medium. On transfer to auxin-free medium, croton tissue formed roots, whereas roots as well as shoots were formed in other genera. In other autotrophic plants, *Santalum*[184] and *Emblica*,[295] the endosperm tissue regenerated by forming embryos.

VI. COMPLETION OF REGENERATION — FORMATION OF PLANTLETS

The shoot buds or embryos formed in vitro can serve for micropropagation when they are able to form plantlets which can be transferred to soil.

A. Rooting of Shoot Bud

For the formation of adventitious roots, individual shoot buds, about 1 cm long, are transferred to a root-induction medium. Low light intensity and a low salt medium are favorable for rooting. Shoots of *Narcissus* formed roots only on half strength MS medium. The response was stimulated by inclusion of auxin, NAA or IBA. However, some plants, like strawberry,[32] *Narcissus*,[294] and *Gladiolus*,[138] readily rooted on hormone-free medium.

Promotion of rooting by riboflavin, in presence of auxin, is reported in *Eucalyptus ficifolia*.[112] Also, phloroglucinol (1,3,5-trihydroxybenzene), a phenolic component from xylem sap of apple trees, promotes rooting of many rosaceous fruit plants.[143,150,151] Promotion of rooting by this compound has been confirmed,[209] but others have found it to be ineffective.[422,423] It is likely that the effect is specific to certain genotypes. Phloroglucinol also promoted IBA-induced rooting.[144,145]

Rooting response of difficult-to-propagate cultivars of apple, Jonathan and Delicious, progressively increased with increasing number of subcultures.[317] Freshly formed shoot buds did not form roots. When shoot buds were subcultured on cytokinin medium to produce successive generation of shoots, after nine subcultures 95% of Jonathan microcuttings formed roots. With Delicious, the percentage of rooting was 21% after 4 subcultures and 79% after 31 subcultures.

Initiation and development of roots is basically promoted by auxin. Cytokinins apparently inhibit this process. The mechanism of interaction between these two groups of plant growth regulators remains unclear. Treatments of petioles of leaf explants of *Phaseolus vulgaris*[40] with indole-butyric acid stimulated root formation on petioles. Immediately after auxin treatment of petioles the level of endogenous cytokinin increased in leaf lamina and decreased in petiole. This can explain the rooting response of petioles.

B. Plantlet from Embryo

Although an embryo differentiating from a callus or in cell suspension culture has a clear radicular end, its maturation into a plantlet has to be distinguished from its initiation. For many species a specific medium is employed for initiation of proliferation and maintenance of a callus. A second medium is employed for the differentiation of embryos, a third medium for the maturation of embryos, and a fourth for their development into plants.

Various factors have been found to affect maturation of embryos in different plants. Although not required for initiation, cytokinin especially favored cotyledon development[9] and maturity[100] of carrot embryos. In umbellifers, *Carum carvi* and *Daucus carota*, the embryos start producing secondary embryos or show precocious germination. These abnormalities can be prevented by a very low level (0.1 μM) of ABA.[6] An effect similar to ABA was possible by 8-azaguanine, an inhibitor of cytokinin activity, in *Carum*.[7] The effect of ABA on embryo maturation has been confirmed in *D. carota*[157] and has also been noted in *Pennisetum americanum*.[389]

Maturity of embryos is also favored by gibberellin in *Citrus sinensis*,[172] *Santalum album*,[184] *Panicum maximum*,[196] and *Zea mays*.[197] Embryos developing from tissue of *Eschscholzia californica*[162] remained dormant which could be overcome either by chilling or gibberellin treatment. Similarly in grape, GA₃ is required for germinaton of mature somatic embryos if chilling is not given. This indicates that endogenous GA₃ may have an important role. It was confirmed in a quantitative study. As the embryos developed, free and water-soluble, GA-like substances increased and chilling at 4°C for 1 week greatly increased the GA contents of embryos.[343] Also, chilling is required at 2 to 4°C for 8 to 10 weeks to overcome apical dormancy of somatic embryos of walnut, *Juglans regia*.[378]

VII. REGENERATION POTENTIAL

Genotypic differences in the ability to form callus and regenerate plants from a callus or a cell suspension culture have been noted for many crop species such as *Brassica*,[19,45,72] *Medicago*,[28,41,231] *Lycopersicon*,[94,95,421] *Lathyrus*,[107] *Cajanus*,[336,337] *Triticum*,[185] *Trifolium*,[50] *Ipomea batatas*,[354] *Hordeum vulgare*,[110] *Zea mays*,[88] and *Oryza*.[2]

Of these studies, in alfalfa it was shown that shoot bud differentiation from callus is controlled by two dominant genes, Rn_1 and Rn_2,[273] and in *Lycopersicon*[94] recessive genes were associated with high shoot-forming capacity. In tomato the regeneration capacity also varied with the developmental and environmental influences.[95]

About the process of regeneration in higher plants it can be summarized that:

1. Regeneration potential is a function of genotype.
2. Tissues from highly regenerative genera undergo differentiation by forming either unipolar structures, roots and shoots, or bipolar structures, embryos. It is possible to evoke or inhibit regeneration, but regulation of regeneration remains to be worked out.
3. Even in highly regenerative types, there is loss of regeneration potential on continuous culture. In this context it is of interest to note that in tissues regenerating to form root and shoot, to begin with, there is loss of shoot-forming potential and this is followed by loss of root-forming potential. Also, in tissues forming embryos, when this potential is lost some cells occasionally form roots.
4. Recalcitrant types, seed legumes and cereals, readily form callus tissue, but this tissue is capable of organization into roots only and fails to form shoots.
5. In some of the recalcitrant types, cereals, if the tissue is raised from meristematic regions, inflorescence, embryo, or leaf meristem, it is possible to have differentiation of bipolar structures, embryos.

However, arising out of this resume are a number of questions:

1. Why is there pattern formation in differentiation? In other words, why do some tissues differentiate by forming embryos and others by forming root and shoot? To be more precise, are there root-, shoot-, and embryo-forming cells?
2. Why is there loss of differentiation potential? Can it be delayed? After having lost, can it be regained?
3. When it is possible to induce formation of roots on shoot buds? Why cannot shoot buds be induced on roots to achieve plantlet formation in recalcitrant types?
4. Do embryos and meristematic tissues have high regeneration potential? In other words, can cell totipotency be correlated with cell origin? To be more precise, are there competent and incompetent cells?
5. Can incompetent cells be made into competent type?

The answers to these questions form the subject matter of this section. Any inference, however, should be taken as tentative because generalization will be possible only when regulation of regeneration is understood.

A. Are there Root-, Shoot-, and Embryo-Forming Cells?

The answer to this question can be affirmative as well as negative.

If one subscribes to the viewpoint that the overall basis of differentiation lies in a control of auxin-cytokinin balance, and in this context, is reviewed the regeneration of recalcitrant types which readily regenerate to produce only roots, then it can be said that there are root-forming cells which are able to express and shoot-forming cells remain repressed. Also, when we consider differentiation in its totality we find that some of the genera are exclusively embryo forming and in other genera are exclusively root and shoot forming. However, even in an embryo-forming genus, such as carrot, different cell populations isolated were either embryogenic or rhizogenic.[314] This finding is consistent with the earlier observation that cell suspension culture of carrot could be reversibly switched between embryogenesis and rhizogenesis[165] by a change in oxygen tension. Low oxygen favored the expression of embryogenic cells and high oxygen of rhizogenic cells.

An evidence in support of the affirmative answer can also be seen in work on regeneration of leaf segments of *Solanum melongena*.[108] The formation of embryos, shoots, and roots was dependent upon different hormone treatments. Individual hormones favored the proliferation of root-, shoot-, and embryo-forming cells. Also, a study of regeneration of leaf segments of *Convolvulus arvensis*[398] indicated that ability to form roots was independent of the ability to form shoots. This indicates that these processes are two independent events and not alternate ones. Possibly, they also do not involve developmentally "plastic" meristemoids.

Loss of differentiation potential is another evidence in favor. In genera showing organogenesis, to begin with, there is loss of shoot formation and it is followed by loss of root formation. This indicates that there are root- and shoot-forming cells.

By contrast, there are evidences which indicate that cells are fully totipotent not only in terms of regeneration capacity, but also in terms of pattern formation. Hormonal and environmental conditions simply modify the expression of differentiation. A cell can form root, shoot, and embryo as well. The microspores of *Nicotiana* readily form embryos, but callus cells of this plant do not form embryos; instead, they undergo organogenesis. However, under high light intensity, callus cells of *N. tabacum* var. Samsun formed embryos.[120] Also, callus cells of *N. sylvestris* formed roots and shoots, but protoplasts of this plant formed embryos.[86,333] Formation of embryos from protoplast-derived tissue is also on record in *N. tabacum*[194] and *Lycopersicon*.[419] Otherwise, tissues from these genera do not form embryos.

However, quite difficult to explain are the instances of formation of embryos as well as shoot and roots, under identical conditions, in a large number of genera. For example, in *Digitalis lutea* and *D. lantana*,[355] cell lines were obtained which were capable of differentiation into (1) embryo, (2) shoot and roots, and (3) roots only. Also, in *Arabidopsis thaliana*[136] is described the formation of roots, shoots, and embryos. On the hypocotyl segments of *Albizzia lebbek* is described the formation of embryos[106] as well as formation of shoots.[381] Similarly, in sweet potato, formation of embryos[193] and shoots[51] have been described. It remains to be resolved whether these differences are due to different genotypes or due to different cultural conditions or even mistaken identity of regenerants.

Control of embryogenesis as well as organogenesis was demonstrated from the callus tissue originating from immature embryos of *Glycine max*.[21] Embryo were formed on MS medium containing 43 μM α-napthalene acetic acid, and organogenesis was possible on medium containing 13.3 μM benzylaminopurine and 0.2 μMα-NAA and four times the concentration of minor salts.

An unequivocal evidence of regeneration via organogenesis as well as embryogenesis was provided in maize tissue,[195] and even biochemical markers could be obtained for shoot- and embryo-forming cultures.[85] When immature embryos of *Zea mays* inbred line, B73, were cultured on MS medium containing 0.5 mg/ℓ 2,4-D and 12% sucrose, a compact white

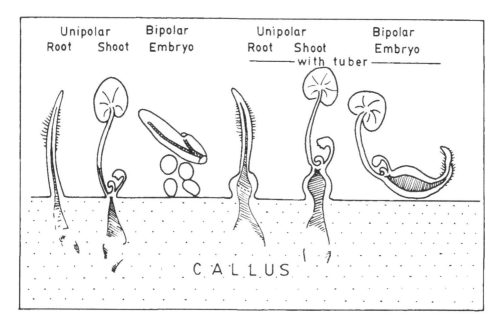

FIGURE 12.   Diagrammatic representation of different types of regenerants from a callus tissue of *Cyclamen*. (Redrawn after Wicart, G., Mouras, A., and Lutz, A., *Protoplasma*, 119, 159, 1984. With permission.)

callus originated from scutellum. When this callus was transferred to same medium, but having 2% sucrose, a green organogenic tissue was selected from compact callus. A sector of this tissue spontaneously became embryogenic. The embryogenic callus could be distinguished by its friability, rapid growth rate, and mucilaginous texture. In addition, isozymes of esterase (EST) and glutamate dehydrogenase (GDH) could be used to distinguish between embryogenic and shoot-forming cultures. The embryogenic callus differentiating in absence of 2,4-D produced GDH isozyme banding pattern characteristic of zygotic embryos, but produced a unique pattern in the presence of 2,4-D. Organogenic callus produced the same pattern as seedling shoot tissue. Undifferentiated callus gave the same pattern as root tissue. An analysis of EST enzyme was also helpful in distinguishing embryogenic, shoot- or root-forming, or undifferentiated tissues. The staining intensities of two of the common EST isozyme bands varied between cell lines; the presence of an additional band in embryogenic cell line was clearly visible.

Of interest in this context is tissue differentiation in *Cyclamen persicum*.[411] Here, there is formation of several kinds of structures: embryos (Figure 12), bipolar tubers (with both shoot and root meristem), unipolar tubers (with only one meristem either for root or shoot), shoot-buds, and roots. In an attempt to explain this diversity of differentiation in terms of a unitary theory, single organogenic pattern, it is hypothesized that from a preembryogenic state all types of regenerants may arise. (1) Somatic embryos formed are bipolar structures. When early tuberization occurs bipolar tubers are formed. Bipolar structures, bipolar tubers or somatic embryos, lack vascular connection with the callus. (2) When one of the meristems aborts early, unipolar structures are formed. Without early tuberization, normal shoot buds and roots are formed. All unipolar structures are connected by vascular strands to the callus. However, in this explanation it is not clear as to how (1) one of the meristems aborts and (2) after the abortion of one of the meristems, how the resulting unipolar structure is able to establish vascular connection with the callus. Therefore, the main question of what determines embryogenesis or organogenesis remains unanswered.

B. Loss of Regeneration Potential

It is a common observation that freshly isolated tissues show a high degree of regeneration.

With continued subculture there is a progressive decline in regeneration potential and ultimately the culture becomes nonmorphogenic. This is true of tissues undergoing organogenesis as well as embryogenesis such as *Nicotiana tabacum*,[225] *Daucus carota*,[338,339] and *Atropa belladonna*.[356] In this respect there are two extremes. In some of the carrot cultures, embryoforming potential is retained for a number of years[279] and in plants like *Atropa*, this potential is rapidly lost[356] within a few subcultures. It is very difficult to comprehend these contrasting situations. The rapid loss of regeneration potential or its retention for a long or short period can be explained in terms of endogenous physiological factors. The capacity for regeneration can be correlated with morphogenic substances carried over in the initial explant and their dilution or degradation in culture. The process can be described to be rapid in those genera which rapidly lose differentiation potential and is transient in others.

The decline and loss of morphogenic potential and cytological abnormalities,[71] hypoploidy, polyploidy, and aneuploidy in tissues in long-term cultures, are parallel features, and it was pointed out that these processes have a causal relationship.[214,225,250] However, it is very difficult to demonstrate a direct relationship between two phenomena, and the most commonly referred evidence in favor of this hypothesis has turned out to be untenable. From a number of tissues showing cytological abnormalities, only diploid plants were regenerated and this was taken as an evidence that only diploid cells are capable of differentiation, and loss of regeneration is accompanied by cytological abnormalities. This is less than correct; plants of various ploidies do regenerate from tissue cultures.

Of the explanations for the difference in differentiation potential of cells in culture, the most attractive is the hypothesis of embryogenic/target/competent cells[123] in culture. According to this proposition, in the primary explant there are specific cells which are capable of differentiation to form embryos/organs in culture. Therefore, a tissue culture is comprised of cells which are competent and incompetent for differentiation. Hence, an active proliferation of competent cells is essential for the expression of differentiation potential, whereas proliferation of incompetent cells would lead to a decline of differentiation potential. The evidences in support of this hypothesis that can be found from literature on embryo formation in carrot cultures are

1. Potential for regeneration can be retained for a prolonged period on medium which does not support differentiation.[278] Normally, in carrot cultures embryo formation begins 4 to 6 weeks after isolation of tissue, and attains a maximum at 15 weeks.
2. After an apparent loss of differentiation potential, this can be restored to some extent on an inductive medium with a high nitrogen level.[278]
3. Slow growth either at low temperature or deficient nutrient medium prolongs differentiation potential.[278]
4. From a population of cells showing a decline in differentiation potential, it is possible to isolate cell lines by plating, which show high differentiation potential.[315] However, even in these cell lines, ultimately a decline in differentiation potential is seen.
5. The decline of differentiation potential is associated with the appearance of cells with altered karyotypes. These cells are, however, selectively favored and ultimately become predominant.[22]
6. Low salt solutions are helpful in preserving the potential of differentiation.[221]

The differentiation potential of a culture can therefore be interpreted in terms of the presence of competent and incompetent cells. Also, there is competition between the two types of cells. Specific cultural conditions favor the growth of a particular cell type. Finally, a few words about slow growth of a tissue which is known to prolong regeneration potential. In shoot-forming tissue of tobacco, sodium sulfate reduced callus growth, but in the absence of Na_2SO_4, this tissue lost regenerative capacity which was, however, retained in the presence of Na_2SO_4.[259] This is the unique effect of Na_2SO_4.

C. Regeneration Potential and Cell Origin

It has been proposed that the nature of explant can be an important factor in obtaining regeneration from cell cultures.[234] Does it imply that not all cells are totipotent? This is not so. It is very likely that cells which are not impaired in differentiation are totipotent.[333] This viewpoint is consistent with early observations that immature embryos of *Cuscuta*[205] formed a tissue which readily differentiated into embryos and more prolific embryo formation was possible when cell suspensions of carrot were derived form embryos.[326] Also, recent successes in regeneration of plants from cereal tissues have been possible due to tissue cultures raised from young embryos or meristematic regions (inflorescence segments or immature basal leaf region). Further, it has been shown that only certain cells are capable of forming a type of tissue which alone is capable of morphogenesis to form embryos.

A strong support to this hypothesis is also possible from work on anther and pollen culture of tobacco. Successul pollen cultures of tobacco leading to the formation of pollen plants were possible only when pollen capable of embryogenesis, which are structurally different from nonembryogenic pollen, were separated and cultured.[271] When two types of pollen were cultured together there was very low response indicating of a competition betweeen competent and incompetent cells.[270]

D. Regeneration Potential and Cell Physiology

That the nature of tissue is a determining factor in its potential for regeneration is relfected in its physiological state. There are a few instances to elaborate this point. During regeneration of leaf explants of *Echeveria*[264] young leaves initiated only roots and older leaves formed shoot buds. English ivy, *Hedera helix,* has juvenile and adult phases.[332] Cultures initiated from juvenile tissue formed roots and shoots,[16] whereas cultures initiated from adult material only produced embryos.[15]

Of interest in this context is the recent work on regeneration in tissue cultures of *Lathyrus sativus.*[107] Regeneration potential in this genus is under genetic control. However, organogenesis was possible in tissues from recalcitrant genotypes by employing physiologically altered explant. When explants were taken from activated lateral bud meristem, consequent to decapitation of apical bud, organogenesis was possible from tissues derived from all genotypes.

E. Lack of Regeneration Potential

At present, the lack of regeneration can be ascribed to either failure of cells in culture to acquire competence under standard conditions or the absence of competent cells. Of interest in this context is a recent study of tissue regeneration from leaf explants of *Convolvulus arvense,*[58] wherein it is shown that it is possible to induce shoots in cultures of genotypes which do not produce them on shoot-inducing medium (SIM) and, conversely, it is possible to induce roots in cultures of genotypes which do not produce roots on root-inducing medium (RIM). Cultures of genotype unable to form roots under standard conditions (RIM) could be made to form roots if they were first cultured on SIM and then transferred to RIM. Similarly, cultures of genotypes unable to form shoots under standard conditions, SIM, could be made to form them if they were first cultured on RIM and then transferred to SIM.

Another interesting aspect of this investigation is information about the location of competent cells. In this material shoots arise from that part of callus which is in contact with the medium and roots arise from top of callus. This is always maintained. The mechanism for this remains obscure, but "physiological grandients" are described in tobacco callus cultures.[286] These two aspects, induction and location of competent cells, need to be looked into in greater detail.

VIII. REGENERATION — AN ANALYSIS

At the very outset, it can be remarked that much is known about factors controlling regeneration, but practically nothing is known about its regulation. Of the various factors, the most significant are hormones, but even about hormonal control, no generalization is possible because various hormones have been shown to promote, inhibit, or be ineffective in many systems. At present the information about regulation of regeneration is fragmentary, and a repertoire of changes are being investigated and correlated to regulate regeneration.

A basic limitation of any information arising out of any one or more of the approaches is that the complex process of differentiation initiates in a number of individual cells which lie scattered in a mass of undifferentiated cells. For an insight into regulation of differentiation different approaches adopted are histro-, cyto-, and biochemical and biophysical.

For an understanding of regulation the two systems commonly employed are tobacco tissue, which undergoes organization by forming monopolar structure roots/shoots, and carrot tissue, which forms embryos. Basically, organogenesis and embryogenesis are two aspects of the same process, differentiation, and hence it is very likely that information from these two systems will serve as a cross check to arrive at conclusions.

A. Nucleic Acids

Cyto-, histo-, and biochemical analyses have revealed that increased nucleic acid and protein syntheses occur during differentiation and are probably necessary for the process. Individual meristemoid cells in tissue of tobacco[369] and proembryonic cells in tissue of carrot[210] have more RNA and protein content during differentiation.

A biochemical transition point in carrot cultures, after 3 days of transfer to an embryo-forming medium, is seen in terms of increased thymidine and leucine incorporation.[101] This is indicative of an increased nucleic acid and protein synthesis in differentiation (Figure 13) and is coincident with the appearance of globular embryos. In another study, an increased rate of protein and RNA synthesis could be detected as early as 2 to 4 hr after transfer to embryo-inducing medium and it continued up to 12 hr of transfer. An attempt to characterize RNA formed during this period revealed that embryogenic cells synthesized more poly-A-containing RNA[298] than nonembryogenic cells, indicating thereby that removal of auxin results in the synthesis of a message. However, the progress of embryogenesis up to globular stage in presence of cordycepin, an inhibitor of polyadenylation, indicates that proteins synthesized during the early stages of embryogenesis are not translation products of poly-A RNA.

B. Proteins

Changes in structural and enzymatic proteins have been implied during differentiation. On application of chloramphenicol to carrot cultures at a concentration which prevented embryogenesis but not cell division, it was concluded that specific proteins are required for differentiation.[339] Also, shoot-forming tobacco callus produced certain specific proteins that are formed in leaves of greenhouse-grown plants.[297]

In a study employing two-dimensional gel electrophoresis, a powerful resolving technique, protein profiles have been followed in embryogenic and nonembryogenic cultures of carrot in the absence and presence of 2,4-D, and two specific proteins have been designated as embryogenic proteins.[334] Irrespective of the presence or absence of auxin, these proteins are synthesized in cells during early periods of growth, but in the presence of 2,4-D these proteins gradually diminish and disappear after 12 days. The presence of these proteins in cells maintained on 2,4-D medium indicates that these proteins are synthesized in the presence of auxin does not allow the embryogenesis to proceed. This reasoning is consistent with the contention that the presence of 2,4-D in the medium allows the proliferation of embryogenic cells, but does not support embryogenesis.

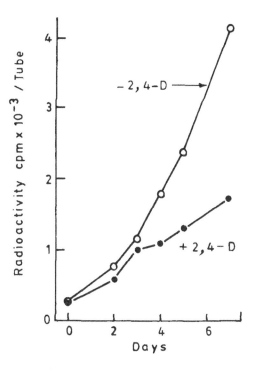

FIGURE 13.   Thymidine incorporation into perchloric acid-insoluble fraction of carrot cells on culture on medium with or without 2,4-D. (Redrawn after Fujimura, T., Komamine, A., and Matsumoto, H., *Physiol. Plant.*, 49, 225, 1980. With permission.)

Auxin receptors are assumed to represent the first site of auxin action in a cell. The role of auxin as a cell division factor and root-forming factor has been resolved in terms of the appearance of a specific membrane-bound, auxin-binding factor during regeneration of roots in tobacco tissue.[201] When tobacco tissue was grown on medium containing α-naphthalene acetic acid and kinetin, three classes of auxin binding proteins could be detected. However, instead of NAA and kinetin, when only 2,4-D was employed one of these sites which is membrane-bound disappeared (Figure 14). On transfer of callus to medium containing NAA and kinetin, this membrane-bound receptor reappeared after 4 to 8 weeks of culture. This reappearance is correlated with the ability of the cells to regenerate roots.

C. Enzymes

It has been implied that a number of enzymes change during differentiation of a tissue. In carrot callus, before the regeneration of roots, the activity of glutamate dehydrogenase increased three-fold and a smaller increase was recorded in respect of aspartate aminotransferase, isocitrate dehydrogenase, and acid phosphatase.[402] Also, a significant increase in polyphenol oxidase[116] was seen prior to and during root formation in carrot tissue. As for shoot formation, a study of nitrogen-assimilating enzymes in shoot- and nonshoot-forming tissues of sugarcane[79] indicated that (1) shoot differentiation is accompanied by peaks in activities of glutamine synthetase, glutamate synthase, and nitrate reductase, whereas glutamate dehydrogenase activity was at its lowest; (2) better mobilization of nitrate occurs in shoot-forming tissue; and (3) glutamine synthetase/glutamate synthase pathway becomes operative prior to differentiation.

Changes in isoperoxidases have been shown during root and shoot formation in tobacco tissue.[289,363] The increase in peroxidase activity prior to organogenesis indicates changes in

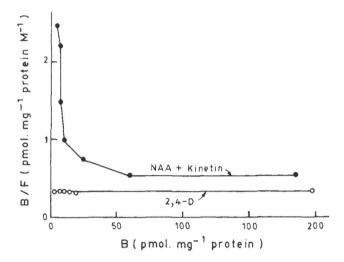

FIGURE 14.   Scatchard plot of NAA binding to isolated membrane fractions from cell suspension cultures of tobacco on medium containing NAA and kinetin or 2,4-D. (From Maan, A. C., Van der Linde, P. C. G., Harkes, P. A. A., and Libbenga, K. R. L., *Planta*, 164, 376, 1985. With permission.)

endogenous auxin in relation to shoot or root initiation. Of the isoperoxidases, anodic bands showed very little activity in shoot-forming tissue as compared to nonshoot-forming tissue. Distinct changes in isoperoxidase patterns (continuous increase in the number and intensity of anodic peroxidases and an increase in intensity of cathodic isoperoxidases up to a maximum followed by a decrease) are indicative of a two-phase requirement of endogenous auxin, a reduction of auxin during induction of shoots, and an increase in auxin during initiation of roots. Also, in the embryogenic cell line of "shamouti" orange there was an upsurge of peroxidase activity concomitant with the appearance of embryos. Interestingly, a new band typical of embryogenic cell line could be seen in isozyme profile.[167]

Proteolytic enzymes are involved in a variety of cellular processes. Recent results indicate that serine protease are involved in biochemical events concerning the initial stage of adventitous bud differentiation on *Torenia* stem segments.[351] This was inferred from experiments in which application of diisopropyl fluorophosphate (DFP), a highly sensitive inhibitor for serine enzymes, strongly inhibited cytokinin-induced adventitious bud intiation in *Torenia*.

Some of the enzyme changes are also covered in the next section and reflect the pool size of metabolites.

D. Metabolites and Metabolism

In tobacco tissue, particularly in shoot-forming regions, there is a heavy accumulation of starch[369,370] prior to the formation of meristematic regions.[365-367] It is followed by a decline during the formation of meristemoids and primordia. The accumulation is interpreted as a result of synthetic activity while reduction, during meristemoid and primordium formation, is ascribed due to enhanced ratio of degradation.[367, 368] Also, in shoot-forming tissue there was continuous incorporation of ^{14}C into starch. Further, it was resolved that there is greater ^{14}C incorporation into shoot-forming than in the growing tissue.[362] More recently,[188,364] it is shown that there is increased α-amylase activity in shoot-forming tissue as compared to nonshoot-forming tissue.[188] There was no effect of medium gibberellic acid on the activity of α-amylase.

It appears that accumulation of reserve storage material is a general feature of cells

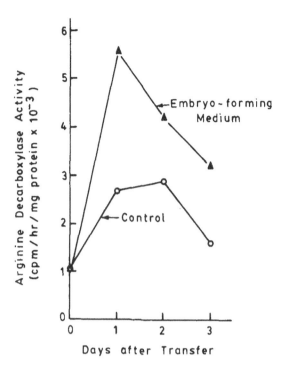

FIGURE 15. Activity of arginine decarboxylase of carrot cells on embryo- and callus-forming media. (From Montague, M. J., Armstrong, T. A., and Jaworski, E. G., *Plant Physiol.*, 63, 34, 1979. With permission.)

undergoing organogenesis and this energy is utilized for the process of differentiation. Also, in embryo-forming cell lines of *Corylus avellana* and *Poulownia tomentosa*, a significant increase in starch content occurs in those cells destined to form embryos and is maintained during embryo formation.[262] This is also true of embryogenesis in *Papaver orientale*.[127]

Changes in polyamine metabolism between embryogenic and nonembryogenic cells of carrot are on record.[216] On the transfer of culture to auxin-free medium, within 24 hr the putrescine level increased twofold and within 6 hr the activity of arginine decarboxylase (Figure 15), required for synthesis of putrescine from arginine, was higher.[215] Also, an increase in ornithine-carbomyltransferase has been reported during the early stage of embryo development in *D. carota*.[13] Further it has been shown that a reduction in the number of embryos formed in carrot tissue occurs after the addition of α-difluoromethylarginine, a specific inhibitor of arginine decarboxylase, and the addition of putrescine could restore the level of embryo formation[89] (Figure 16).

E. Osmolarity and Communication

The role of biophysical events in differentiation is the least understood. It is quite reasonable to assume that these events are equally important along with biochemical changes. Some of the changes implied are changes in membrane permeability and the consequent osmolarity and cell communication.

In tobacco tissue it appears that there is specific osmotic component for shoot formation because only one third of the sucrose in the medium is sufficient for it and it could be replaced by a nonmetabolizable sugar, mannitol.[42] Further, shoot-forming tissue maintained greater water, osmotic potentials, and pressure than the unorganized tissue.[43] These differences in potentials were observed by day 2 of culture and were maximum by day 6, a time

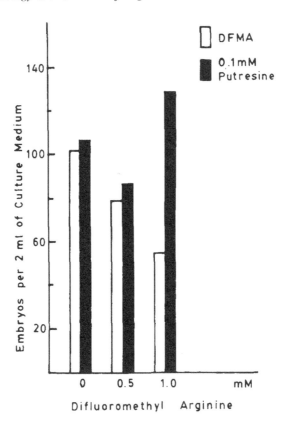

FIGURE 16.   Effect of α-difluoromethylarginine, a specific inhibitor of decarboxylase activity on embryo formation by carrot cells. (From Feirer, R. P., Mignon, G., and Litray, J. D., *Science*, 223, 1433, 1984. With permission.)

for appearance of changes leading to the formation of meristemoids, and were maintained throughout the culture period. A possible consequence of increased osmotic potential is enhancement of mitochondrial activity. A study of isolated mitochondria (Figure 17) showed that shoot-forming tissue had higher and more efficient respiration than nonshoot-forming tissue.[44] The role of osmotic potential is to be looked into in greater detail because in potato, osmotic conditions for shoot-bud formation are quite different from root formation.[166,301]

As for cell integration, it is widely believed that plasmodesmata play an important role in cell communication. In lower plants plasmolysis is frequently used to initiate regeneration. Also, in *D. carota* frequency of embryo formation could be increased to threefold by a 45-min plasmolysis[406] treatment employing 1 *M* sucrose. It is very likely that plasmolysis disrupts cellular interconnections allowing more cells to develop independently and expressing their totipotency through embryo formation. This concept of the developmental role of cell isolation by snapping of plasmodesmata is consistent with the subcellular findings. In many genera embryo-sac-forming meiocyte is isolated from adjacent cells by a thick wall lacking plasmodesmata, and this is also the situation with developing zygote and embryo.[293]

An indirect evidence for the importance of cell communication in cell differentiation is seen in the electrophysiological property[342] of tobacco cells to their regenerative ability. Cell clusters in high density which readily formed leaf primordia on medium containing kinetin had higher membrane potential than low density cell fraction in which leaf primordia rarely differentiated. That the electrical properties of cells have something to do with their differentiation was discovered when the number of shoots formed on tobacco callus was increased

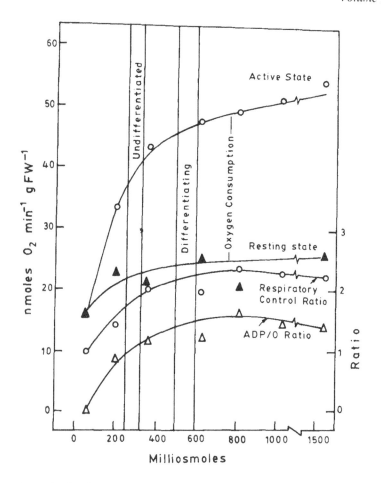

FIGURE 17. Organogenesis in relation to metabolism, activity of mitochondria, isolated from shoot-forming tobacco callus, at different osmolarities. Bars represent osmolarity levels of undifferentiated and tissue undergoing differentiation. (From Thorpe, T. A., in *Genetic Engineering Application to Agriculture*, Owens, L. D., Ed., Rowman and Allanheld, London, 1983, 285. With permission.)

by up to fivefold by passing a very weak electric current between the culture and the medium.[111] This unexpected finding has been extended to other systems such as wheat and can be of practical application in inducing regeneration in tissues which are recalcitrant. A follow up of this work is likely to be rewarding.

F. Gene Expression

Of central importance in the process of differentiation are the molecular events leading to the differential activation of genes. For regulation of gene expression, histones[135] as well as nonhistone chromosomal proteins[323] have been implied. Changes in chromosomal proteins recorded during embryo formation in carrot[87,101,115] tissue have been interpreted to facilitate gene expression.

G. Polarity

Polarity is described as a prerequisite for organized growth,[303] and it is hypothesized[333] that if polarity is established within the first cell, embryo is formed and if it is established within a group of cells (meristemoid), unipolar structure, root, or shoot is formed. The establishment of polarity in an individual cell is manifested by an initial asymmetric division.

Embryogenesis in isolated cells of carrot is also preceded by an asymmetric cell division.[10] However, it remains to be seen whether this first unequal division controls embryo formation, because embryos appear much later, only after the formation of a multicellular tissue.

During organogenesis polarity shifts occur. The histological criterion for polarity is the direction of cell division, and the biophysical criterion is the reorientation of cellulose microfibrils.[114] Unfortunately, it has not been possible to investigate these aspects in a callus tissue.

Of interest in this context is the recent work on direct induction of embryos on young zygotic embryos of *Trifolium* in response to applied benzylaminopurine.[204] The first sign of embryo induction on hypocotyl axis was a change from regular anticlinal division to irregular periclinal and oblique division. The effect of BAP is interpreted in terms of not the entry of cells into mitosis per se, but an alteration in cell polarity and division plane,[413] leading to the formation of embryos.

IX. CORRELATION

The concept of cell totipotency in plants is very well established, but least understood. In a differentiating cell the events occurring can be classified as genetical, physiological, biochemical, biophysical, and molecular. Of the physiological events which trigger cell differentiation, only the hormonal control is well worked out, but cannot be correlated for want of knowledge about endogenous hormones and their levels. Also, for a full appreciation of the role of hormones, the other aspect which needs to be studied is cell communication, because in a callus tissue a responding cell lying scattered among nonresponding cells is like a needle in a haystack. It remains to be ascertained whether cell isolation as well as cell communication are essential features of cell differentiation in plants. As for biophysical events, changes in membrane permeability and osmolarity and polarity of cells are the likely factors controlling differentiation, but without much information. The biochemical events which precede organized development imply a shift in metabolites and metabolism that lead to changes in the content and spectra of structural and enzymatic proteins. However, nothing is known about molecular events which regulate differentiation in cell cultures.

REFERENCES

1. **Abe, T. and Futsuhara, Y.,** Efficient plant regeneration by somatic embryogenesis from root callus tissue of rice *(Oryza sativa)*, *J. Plant Physiol.*, 121, 111, 1985.
2. **Abe, T. and Futsuhara, Y.,** Genotypic variability for callus formation and plant regeneration in rice *(Oryza sativa)*, *Theor. Appl. Genet.*, 72, 3, 1986.
3. **Abo El. Nil, M. M.,** Organogenesis and embryogenesis in callus cultures of garlic *(Allium sativum)*, *Plant Sci. Lett.*, 9, 259, 1977.
4. **Aghion-Prat, D.,** Neoformation de fleurs in vitro ches *Nicotiana tabacum*, *Physiol. Veg.*, 3, 229, 1965.
5. **Ahn, B. J., Huang, F. H., and King, J. W.,** Plant regeneration through somatic embryogenesis in common bermuda grass tissue culture, *Crop. Sci.*, 25, 1107, 1985.
6. **Ammirato, P. V.,** The effect of abscisic acid on the development of somatic embryos from cells of caraway *(Carum carva)*, *Bot. Gaz.*, 135, 328, 1974.
7. **Ammirato, P. V.,** The effect of 8-azaguanine on the development of somatic embryos from cultured caraway cells, *Plant Physiol.*, Suppl. 61, 46, 1978.
8. **Ammirato, P. V.,** Somatic embryogenesis and plantlet development in suspension cultures of medicinal yam, *Dioscorea floribunda, Am. J. Bot.*, Suppl. 65, 89, 1978.
9. **Ammirato, P. V. and Steward, F. C.,** Some effects of the environment on the development of embryos from cultured free cells, *Bot. Gaz.*, 132, 149, 1971.
10. **Backs-Husseman, D. and Reinert, J.,** Embryobildung durch isolierte Einzelzellen aus Gewebekulturen von *Daucus carota, Protoplasma*, 70, 49, 1970.

11. **Badila, P., Lauzac, M., and Paulet, P.,** The characteristics of light in floral induction in vitro in *Cichorium intybus.* The possible role of phytochrome, *Physiol. Plant.,* 65, 305, 1985.

12. **Bagga, S., Rajasekar, V. K., Guha-Mukerjee, S., and Sopory, S. K.,** Enhancement in the formation of shoot initials by phytochrome in stem callus cultures of *Brassica oleracea* var. Botrytes, *Plant Sci.,* 38, 61, 1985.

13. **Baker, S. R. and Yon, R. J.,** Characterization of carbamoyl-transferase from cultured carrot cells of low embryogenic potential, *Phytochemistry,* 22, 2171, 1983.

14. **Ball, E. A.,** Regeneration from isolated portions of the shoot apex of *Trachymena coerulea, Ann. Bot.,* 45, 103, 1980.

15. **Banks, M. S.,** Plant regeneration from callus of two growth phases of English ivy, *Hedera helix, Z. Pflanzenphysiol.,* 92, 349, 1979.

16. **Banks, M. S. and Hackett, W. P.,** Differentiation from *Hedera helix* callus, *Plant Physiol,* Suppl. 61, 45, 1978.

17. **Bapat, P. S. and Rao, V. A.,** Shoot apical meristem culture of *Pharbitis nil, Plant Sci. Lett.,* 10, 327, 1977.

18. **Barlass, M. and Skene, K. G. M.,** Studies on the fragmented shoot apex of grape vine. II. Factors affecting growth and differentiation in vitro, *J. Exp. Bot.,* 31, 489, 1980.

19. **Baroncelli, S., Buiatti, M., and Binnici, A.,** Genetics of growth and differentiation in vitro of *Brassica oleracea* var. *botrytis.* Differences between six inbred lines, *Z. Pflanzenzuecht.,* 70, 99, 1973.

20. **Baruch, R. and Quak, F.,** Virus-free plants of *Iris* ''Wedgewood'' obtained by meristem culture, *Neth. J. Plant Pathol.,* 72, 270, 1966.

21. **Barwale, U. B., Kerns, H. R., and Widholm, J. M.,** Plant regeneration from callus cultures of several soybean genotypes via embryogenesis and organogenesis, *Planta,* 167, 473, 1986.

22. **Bayliss, M. W.,** Factors affecting the frequency of tetraploid cells in a predominantly diploid suspension culture of *Daucus carota, Protoplasma,* 92, 109, 1977.

23. **Behki, R. M. and Lesley, S. M.,** In vitro plant regeneration from leaf explants of *Lycopersicon esculentum, Cap. J. Bot.,* 54, 2409, 1976.

24. **Bhargawa, S. and Chandra, N.,** In vitro differentiation in callus cultures of moth bean, *Vigna aconitifolia, Plant Cell Rep.,* 2, 47, 1983.

25. **Bhattacharya, P. and Sen, S. K.,** Potentiality of leaf sheath cells for regeneration of rice *(Oryza sativa)* plants, *Theor. Appl. Genet.,* 58, 87, 1980.

26. **Bhojwani, S. S. and Johri, B. M.,** Cytokinin-induced shoot-bud differentiation in mature endosperm of *Scurrula pulverulenta, Z. Pflanzenphysiol.,* 63, 269, 1970.

27. **Bhojwani, S. S. and Johri, B. M.,** Morphogenic studies on cultured mature endosperm of *Croton bonplandianum, New Phytol.,* 70, 761, 1971.

28. **Bingham, E. T., Hurley, L. V., Kaatz, D. M., and Saunders, J. W.,** Breeding alfalfa which regenerates from callus tissue in culture, *Crop. Sci.,* 15, 719, 1975.

29. **Bonnett, H. T. and Torrey, J. G.,** Comparative anatomy of endogenous bud and lateral root formation in *Convolvulus arvensis* roots cultured in vitro, *Am. J. Bot.,* 53, 496, 1966.

30. **Bouniols, A.,** In vitro neoformation of inflorescence buds from *Cichorium intybus* root fragments, influence of tissue hydration, consequence of amino acid composition, *Plant Sci. Lett.,* 2, 363, 1974.

31. **Bouniols, A. and Margara, J.,** *Ann. Physiol. Veg.,* 10, 69, 1968.

32. **Boxus, P.,** The production of strawberry plants by in vitro micropropagation, *J. Hortic. Sci.,* 49, 209, 1974.

33. **Boyes, C. J. and Vasil, I. K.,** Plant regeneration by somatic embryogenesis from cultured young inflorescence of *Sorghum arudinaceum, Plant Sci. Lett.,* 35, 153, 1984.

34. **Brants, D. H.,** A revised medium for freesia meristem culture, *Neth. J. Plant Pathol.,* 74, 120, 1968.

35. **Brants, D. H. and Vermeulen, H.,** Production of virus-free freesias by means of meristem culture, *Neth. J. Plant Pathol.,* 71, 25, 1965.

36. **Branton, R. L. and Blake, J.,** Development of organized structures in callus derived from explants of *Cocos nucifera, Ann. Bot.,* 52, 673, 1983.

37. **Broertjes, C., Haccius, B., and Weidlich, S.,** Adventitious bud formation on isolated leaves and its significance in mutation breeding, *Euphytica,* 17, 321, 1968.

38. **Brossard-Chriqui, D. and Tripathi, B. K.,** Comparison des aptitudes morphogenetiques des etamines fertiles ou steriles d' *Actinida chinensis* cultivees in vitro, *Can. J. Bot.,* 62, 1940, 1984.

39. **Brawley, S. H., Wetherell, D. F., and Robinson, K. R.,** Electrical polarity in embryos of wild carrot precedes cotyledon differentiation, *Proc. Natl. Acad. Sci. U.S.A.,* 81, 6064, 1984.

40. **Bridglall, S. S. and Van Staden, J.,** Effects of auxin on rooting and endogenous cytokinin levels in leaf cuttings of *Phaseolus vulgaris, J. Plant Physiol.,* 117, 287, 1985.

41. **Brown, D. C. W. and Atanassov, A.,** Role of genetic background in somatic embryogenesis in *Medicago, Plant Cell Tissue Organ Culture,* 4, 111, 1985.

42. **Brown, D. C. W., Leung, D. W. M., and Thorpe, T. A.,** Osmotic requirements for shoot formation in tobacco callus, *Physiol. Plant.,* 46, 36, 1979.
43. **Brown, D. C. W. and Thorpe, T.,** Changes in water potential and its components during shoot formation in tobacco callus, *Physiol. Plant.,* 49, 83, 1980.
44. **Brown, D. C. W. and Thorpe, T. A.,** Mitochondrial activity during shoot formation and growth in tobacco callus, *Physiol. Plant.,* 54, 125, 1982.
45. **Buiatti, M., Barnocelli, S., Binnici, A., Pagliai, M., and Tesi, R.,** Genetics of growth and differentiation 'in vitro' of *Brassica oleracea* var. *botrytis* II. An in vitro and in vivo diallele cross, *Z. Pflanzenzuecht.,* 72, 269, 1974.
46. **Burg, S. P. and Burg, E. A.,** Auxin-stimulated ethylene formation: its relationship to auxin inhibited growth, root geotropism and other plant process, in *Biochemistry and Physiology of Plant Growth Substances,* Wightman, F. and Setterfield, G., Eds., Runge Press, Ottawa, 1968, 1275.
47. **Butcher, D. and Street, H. E.,** Excised root culture, *Biol. Rev.,* 30, 513, 1964.
48. **Button, J. and Bornman, C. H.,** Development of nucellar plants from unpollinated and unfertilized ovules of Washington Naval Orange in vitro, *J. S. Afr. Bot.,* 37, 127, 1971.
49. **Campbell, A. I.,** Apple virus inactivation by host therapy and tip propagation, *Nature (London),* 195, 520, 1962.
50. **Campbell, C. T. and Tomes, D. T.,** Establishment and multiplication of red clover plants by in vitro shoot tip culture, *Plant Cell Tissue Organ Culture,* 3, 49, 1984.
51. **Carswell, G. K. and Locy, R. D.,** Root and shoot initiation by leaf, stem, and storage root explants of sweet potato, *Plant Cell Tissue and Organ Culture,* 3, 229, 1984.
52. **Chang, W. C. and Chiu, P. L.,** *J. Chin. Soc. Hortic. Sci.,* 22, 73, 1976.
53. **Chaturvedi, H. C., Misra, P., and Sharma, M.,** In vitro multiplication of *Rosmarinus officinalis, Z. Pflanzenphysiol.,* 113, 301, 1984.
54. **Chaturvedi, H. C. and Sinha, M.,** Mass clonal propagation of *Solanum khasianum* through tissue culture, *J. Exp. Biol.,* 17, 153, 1979.
55. **Chen, T. H., Lain, L., and Chen, S. C.,** Somatic embryogenesis and plant regeneration from cultured young inflorescence of *Oryza sativa, Plant Cell Tissue Organ Culture,* 4, 51, 1985.
56. **Chlyah-Arnasson, A. and Tran Thanh Van, K.,** Budding capacity of undetached *Begonia rex* leaves, *Nature (London),* 218, 493, 1968.
57. **Christianson, M. L. and Warnick, D. A.,** Phenocritical times in the process of in vitro shoot organogenesis, *Dev. Biol.,* 101, 382, 1984.
58. **Christianson, M. L. and Warnick, D. A.,** Temporal requirement for phytohormone balance in the control of organogenesis in vitro, *Dev. Biol.,* 112, 494, 1985.
59. **Coleman, W. K. and Greyson, R. I.,** Analysis of root formation in leaf discs of *Lycopersicon esculentum* cultured in vitro, *Ann. Bot.,* 41, 307, 1977.
60. **Coleman, W. K. and Greyson, R. I.,** Promotion of root initiation by gibberellic acid in leaf discs of tomato *(Lycopersicon esculentum)* cultured in vitro, *New Phytol.,* 78, 47, 1977.
61. **Coleman, W. K., Huxter, T. J., Reid, D. M., and Thorpe, T. A.,** Ethylene as an endogenous inhibitor of root regeneration in tomato leaf discs in vitro, *Physiol. Plant.,* 48, 519, 1980.
62. **Collins, G. B. and Phillips, G. C.,** In vitro tissue culture and plant regeneration in *Trifolium pratense,* in Proc. U.S.A.-N.S.F. and France C.N.R.S. Semin. on Plant Regeneration, Paris, 1980 (Abstr.).
63. **Conger, B. V.,** *Cloning Agricultural Plants via In Vitro Techniques,* CRC Press, Boca Raton, Fla., 1981.
64. **Conger, B. V., Hanning, G. E., Gray, D. J., and McDaniel, J. K.,** Direct embryogenesis from mesophyll cells of orchard grass, *Science,* 221, 850, 1983.
65. **Constabel, F., Miller, R.A., and Gamborg, O.,** Histological studies on embryos produced from cell cultures of *Bromus inermis, Can. J. Bot.,* 49, 1415, 1971.
66. **Cousson, A. and Trant Thanh Van, K.,** In vitro control of de novo flower differentiation from tobacco thin cell layer on a liquid medium, *Physiol. Plant.,* 51, 77, 1980.
67. **Cousson, A. and Tranh Thanh Van, K.,** Light and sugar-mediated control of direct de novo flower differentiation from tobacco cell layers, *Plant Physiol.,* 72, 33, 1983.
68. **Croes, A. F., Creemers-Melennar, T., Van den Ende, G., Kemp, A., and Barendse, G. W. M.,** Tissue age as an endogenous factor controlling in vitro bud formation on explants from the inflorescence of *Nicotiana tabacum, J. Exp. Bot.,* 36, 1771, 1985.
69. **Crooks, D. M.,** Histological and regenerative studies on the flax seedlings, *Bot. Gaz.,* 95, 209, 1933.
70. **Dale, P. J.,** Embryoids from cultured immature embryos of *Lolium multiflorum, Z. Pflanzenphysiol.,* 100, 73, 1980.
71. **D'Amato, F.,** Cytogenetics of differentiation in tissue and cell cultures, in *Applied and Fundamental Aspects of Plant Cell Tissue and Organ Culture,* Reinert, J. and Bajaj, Y. P. S., Eds., Springer-Verlag, Berlin, 1977, 343.
72. **Dietert, M. F., Barron, S. A., and Yoder, O. C.,** Effects of genotype on in vitro culture in the genus *Brassica, Plant Sci. Lett.,* 26, 233, 1982.

73. **Dore, J.,** Physiology of regeneration in cormophytes, in *Encyclopaedia of Plant Physiology,* XV/2, Springer-Verlag, Basel, 1965, 1.

74. **Doerschug, M. R. and Miller, C. O.,** Chemical control of adventitious organ formation in *Lactuca Sativa* explants, *Am. J. Bot.,* 54, 410, 1967.

75. **Dougall, D. K. and Shimbayashi, K.,** Factors affecting growth of tobacco callus tissue and its incorporation of tyrosine, *Plant Physiol.,* 35, 396, 1960.

76. **Douglas, G. C.,** Formation of adventitious buds in stem internodes of populus hybrid TT 32 cultured in vitro. Effects of sucrose, zeatin, IAA and ABA, *J. Plant Physiol.,* 121, 225, 1985.

77. **Duncan, D. R., Williams, M. F., Zehr, B. E., and Widholm, J. M.,** The production of callus capable of plant regeneration from immature embryos of numerous *Zea mays* genotypes, *Planta,* 165, 322, 1985.

78. **Dunwell, J. M.,** In vitro regeneration from excised leaf discs of three *Brassica* species, *J. Exp. Bot.,* 129, 789, 1981.

79. **Dwivedi, U. N., Khan, B. M., Rawal, S. K., and Mascarenhas, A. F.,** Biochemical aspects of shoot differentiation in sugarcane callus. I. Nitrogen assimilating enzymes, *J. Plant Physiol.,* 117, 7, 1985.

80. **Earle, E. D. and Langhans, R. W.,** Propagation of *Chrysanthemum* in vitro. II. Production, growth and flowering of plantlets from tissue cultures, *J. Am. Soc. Hortic. Sci.,* 99, 352, 1974.

81. **Ende Van Den, G., Barendse, G. W. M., Kemp, A., and Croes, A. F.,** The role of glucose on flower-bud formation in thin layer tissue cultures of *Nicotiana tabacum, J. Exp. Bot.,* 35, 1853, 1984.

82. **Ende Van Den G., Croes, A. F., Kemp, A., Barendse, G. W. M., and Kroh, M.,** Floral morphogenesis in thin layer tissue cultures of *Nicotiana tabacum, Physiol. Plant.,* 62, 83, 1984.

83. **Ernst, D. and Oesterhelt, D.,** Changes in cytokinin nucleotides in an anise cell culture *(Pinpenella anisum)* during growth and embryogenesis, *Plant Cell Rep.,* 4, 140, 1985.

84. **Espinoza, N. O. and Dodds, J. H.,** Adventitious shoot formation on cultured potato root, *Plant Sci.,* 41, 121, 1985.

85. **Everett, N. P., Wach, M. J., and Ashworth, D. J.,** Biochemical markers of embryogenesis in tissue cultures of the maize inbred 73, *Plant Sci.,* 41, 133, 1985.

86. **Facciotti, D. and Pilet, P. E.,** Plants and embryoids from haploid *Nicotiana sylvestris* protoplasts, *Plant Sci. Lett.,* 15, 1, 1979.

87. **Facciotti, D., Pilet, P. E., and Matsumoto, H.,** Changes in chromosomal proteins during early stages of synchronized embryogenesis in a carrot cell suspension culture, *Z. Pflanzenphysiol.,* 102, 293, 1981.

88. **Fahey, J. W., Reed, J. N., Reddy, T. L., and Pace, G. M.,** Somatic embryogenesis from three commercially important inbreds of *Zea mays, Plant Cell Rep.,* 5, 35, 1986.

89. **Feirer, R. P., Mignon, G., and Litray, J. D.,** Arginine decarboxylase and polyamines required for embryogenesis in the wild carrot, *Science,* 223, 1433, 1984.

90. **Feng, K. A. and Linck, A. J.,** Effects of *N* -1-naphthylpthalamic acid on the growth and bud formation of tobacco callus grown in vitro, *Plant Cell Physiol.,* 11, 589, 1970.

91. **Feung, C., Hamilton, R. H., and Mumma, R. O.,** Metabolism of 2,4-D. Identification of metabolites in rice root callus tissue cultures, *J. Agric. Food Chem.,* 24, 1013, 1976.

92. **Feung, C., Hamilton, R. H., and Mumma, R. O.,** Metabolism of indole-3-acetic acid, *Plant Physiol.,* 59, 91, 1977.

93. **Finer, J. J. and Smith, R. H.,** Initiation of callus and somatic embryos from explants of mature cotton, *Plant Cell Rep.,* 3, 41, 1984.

94. **Frankenberger, E. A., Hasegawa, P. M., and Tigchelaar, E. C.,** Diallele analysis of shoot-forming capacity among selected tomato genotypes, *Z. Pflanzenphysiol.,* 102, 233, 1980.

95. **Frankenberger, E. A., Hasegawa, P. M., and Tigchelaar, E. C.,** Influence of environment and developmental state on the shoot-forming capacity of tomato genotypes, *Z. Pflanzenphysiol.,* 102, 221, 1980.

96. **Fridborg, G. and Eriksson, T.,** Partial reversal by cytokinin and (2 chloroethyl) -trimethylammonium chloride of near-ultraviolet-inhibited growth and morphogenesis in callus cultures, *Physiol. Plant.,* 34, 162, 1975.

97. **Fujimura, T. and Komamine, A.,** Effects of various growth regulators on the embryogenesis in a carrot cell suspension culture, *Plant Sci. Lett.,* 5, 359, 1975.

98. **Fujimura, T. and Komamine, A.,** Involvement of endogenous auxin in somatic embryogenesis in a carrot cell suspension culture, *Plant Sci. Lett.,* 5, 359, 1979.

99. **Fujimura, T. and Komamine, A.,** Mode of action of 2,4-D and zeatin on somatic embryogenesis in a carrot cell suspension culture, *Z. Pflanzenphysiol.,* 99, 1, 1980.

100. **Fujimura, T. and Komamine, A.,** The serial observation of embryogenesis in a carrot cell suspension culture, *New Phytol.,* 86, 213, 1980.

101. **Fujimura, T., Komamine, A., and Matsumoto, H.,** Aspects of DNA, RNA and protein synthesis during somatic embryogenesis in a carrot cell culture, *Physiol. Plant.,* 49, 255, 1980.

102. **Furuya, M. and Torrey, J. G.,** The reversible inhibition by red and far-red light of auxin-induced lateral root initiation in isolated pea roots, *Plant Physiol.,* 39, 987, 1964.

103. **Gamborg, O., Constabel, F., and Miller, R. A.,** Embryogenesis and production of albino plants from cell cultures of *Bromus inermis, Planta,* 95, 355, 1970.

104. **Gamborg, O., Constabel, F., and Shyluk, J. P.,** Organogenesis in callus from shoot apices of *Pisum sativum, Physiol. Plant.,* 30, 125, 1974.

105. **Gautheret, R. J.,** *Une Voie nouvelle en Biologie vegetale, La culture de Tissus Vegetaux,* Gallimard, Paris, 1945.

106. **Gharyal, P. K. and Maheshwari, S. C.,** In vitro differentiation of somatic embryoids in a leguminous tree *Albizzia lebbeck, Naturwissenschaften,* 68, 379, 1981.

107. **Gharyal, P. K. and Maheshwari, S. C.,** Genetic and physiological influences on differentiation in tissue cultures of a legume *Lathyrus sativus, Theor. Appl. Genet.,* 66, 123, 1983.

108. **Gleddie, S., Keller, W., and Setterfield, G.,** Somatic embryogenesis and plant regeneration from leaf explants and cell suspension of *Solanum melongena, Can. J. Bot.,* 61, 656, 1983.

109. **Godbole, D. A., Kunachgi, M. N., Potdar, U. A., Krishnamurthy, K. V., and Mascarenhas, A. F.,** Studies on a drought resistant legume: the moth bean, *Vigna aconitifolia.* II. Morphogenetic studies, *Plant Cell Rep.,* 3, 75, 1984.

110. **Goldstein, C. S. and Kronstad, W. F.,** Tissue culture and plant regeneration from immature embryo explants of Barley, *Hordeum vulgare, Theor. Appl. Genet.,* 71, 631, 1986.

111. **Goldsworthy, A. and Rathore, K. S.,** The electrical control of growth in plant tissue cultures: the polar transport of auxin, *J. Exp. Bot.,* 36, 1134, 1985.

112. **Gorst, J. R. and De Fossard, R. A.,** Riboflavin and root morphogenesis in *Eucalyptus,* in *Plant Cell Cultures: Results and Perspectives,* Sala, F., Parisi, B., Cella, R., and Cifferi, O., Eds., Elsevier, Amsterdam, 1980, 271.

113. **Grant, J.,** Plant regeneration from cotyledonary tissue of *Glycine canescens,* a perennial wild relative of soybean, *Plant Cell Tissue Organ Culture,* 3, 169, 1984.

114. **Green, P. B. and Lang, J. M.,** Toward a biophysical theory of organogenesis. Birefringence observations on regenerating leaves in the succulent *Graptopetalum paraguayense, Planta,* 151, 413, 1981.

115. **Gregor, D., Reinert, J., and Matsumoto, H.,** Changes in chromosomal proteins from embryo-induced carrot cells, *Plant Cell Physiol.,* 15, 875, 1974.

116. **Habaguchi, K.,** Purification and some properties of polyphenol oxidase from root forming carrot callus tissues, *Plant Cell Physiol.,* 20, 9, 1979.

117. **Haccius, B.,** Zur derzeitigen situation der Angiospermen-Embryologie, *Bot. Jahrb.,* 91, 309, 1971.

118. **Haccius, B.,** Question of unicellular origin of non-zygotic embryos in callus cultures, *Phytomorphology,* 28, 74, 1978.

119. **Haccius, B. and Bhandari, N. N.,** Zur Frage der ''Befestigung'' junger pollen embryonen von *Nicotiana tabacum* an der Antheren Wand, *Beitr. Biol. Pflanz.,* 51, 53, 1975.

120. **Haccius, B. and Lakshmanan, K. K.,** Adventivembryonen aus *Nicotiana* Kallus, der bei hohen Lichtintensitaten Kultiviert wurde, *Planta,* 65, 102, 1965.

121. **Haddon, L. and Northcote, D. H.,** The effect of growth conditions and origin of tissue on the ploidy and morphogenetic potential of tissue cultures of bean *(Phaseolus vulgaris), J. Exp. Bot.,* 27, 1031, 1976.

122. **Halperin, W.,** Alternative morphogenetic events in cell suspension, *Am. J. Bot.,* 53, 443, 1966.

123. **Halperin, W.,** Population density effects on embryogenesis in carrot cell cultures, *Exp. Cell Res.,* 48, 170, 1967.

124. **Halperin, W.,** Embryos from somatic plant cells, in *Control Mechanisms in the Expression of Cellular Phenotypes,* Padykula, H. A., Ed., Academic Press, New York, 1970, 169.

125. **Halperin, W. and Wetherell, D. F.,** Adventive embryony in tissue cultures of the wild carrot, *Daucus carota, Am. J. Bot.,* 51, 274, 1964.

126. **Halperin, W. and Wetherell, D. F.,** Ontogeny of adventive embryony in tissue cultures of wild carrot, *Science,* 147, 756, 1965.

127. **Hara, S., Falk, H., and Kleinig, H.,** Starch and triacyl-glycerol metabolism related to somatic embryogenesis in *Papaver orientale* tissue cultures, *Planta,* 164, 303, 1985.

128. **Hartman, R. D.,** Dasheen mosaic virus and other phytopathogens eliminated from *Caladium,* taro and coco yam by culture of shoot tips, *Phytopathology,* 64, 237, 1974.

129. **Haydu, Z. and Vasil, I. K.,** Somatic embryogenesis and plant regeneration from leaf tissues and anthers of *Pennisetum purpureum, Theor. Appl. Genet.,* 59, 269, 1981.

130. **Heidi, O. M.,** Stimulation of adventitious bud formation in Begonia leaves by absicisic acid, *Nature (London),* 219, 960, 1968.

131. **Heyser, J. W., Dykes, T. A., DeMott, K. J., and Nabors, M. W.,** High frequency long term regeneration of rice from callus cultures, *Plant Sci. Lett.,* 29, 175, 1983.

132. **Holmes, F. O.,** Elimination of spotted wild from *Dahlia, Phytopathology,* 38, 314, 1948.

133. **Hu, C. Y. and Sussex, I. M.,** In vitro development of embryoids on cotyledons of *Ilex aquifolium, Phytomorphology,* 23, 103, 1971.

134. **Hu, C. Y., Ochs, J. D., and Mancini, F. M.**, Further observations on *Ilex* embryoid production, *Z. Pflanzenphysiol.*, 89, 41, 1978.

135. **Huang, R. C. and Bonner, J.**, Histone, a suppressor of chromosomal ribonucleic acid synthesis, *Proc. Natl. Acad. Sci. U.S.A.*, 48, 1216, 1952.

136. **Huang, B. C. and Yeoman, M. M.**, Callus proliferation and morphogenesis in tissue cultures of *Arabidopsis thaliana, Plant Sci. Lett.*, 33, 353, 1984.

137. **Hussey, G.**, Plantlet regeneration from callus and parent tissue in *Ornithogalum thyrsoides, J. Exp. Bot.*, 27, 375, 1976.

138. **Hussey, G.**, In vitro propagation of some members of Liliaceae, Iridaceae and Amaryllidaceae, *Acta Hortic.*, 78, 303, 1977.

139. **Huxter, T. J., Reid, P. M., and Thorpe, T. A.**, Shoot initiation in light and dark grown tobacco callus. The role of ethylene, *Physiol. Plant.*, 53, 319, 1981.

140. **Isogai, Y., Shudo, K., and Okamoto, T.**, Effect of *N*-phenyl-*N*-(4-pyridyl) urea on shoot formation in tobacco pith disc and callus cultured in vitro, *Plant Cell Physiol.*, 17, 591, 1976.

141. **Jacobsen, H. J. and Kysley, W.**, Induction of somatic embryos in pea, *Pisum sativum, Plant Cell Tissue Organ Culture*, 3, 319, 1984.

142. **Jacquoit, C.**, Action du mesoinositol et de l'adenine sur la formation de bourgeons par le tissu cambial, *C. R. Acad. Sci.*, 233, 815, 1951.

143. **James, D. J.**, The role of auxins and phloroglucinol in adventitious root formation in *Rubus* and *Fragaria* grown in vitro, *J. Hortic. Sci.*, 54, 273, 1979.

144. **James, D. J.**, Shoot and root initiation in vitro in the apple root stock M9 and the promotive effects of phloroglucinol, *J. Hortic. Sci.*, 56, 15, 1981.

145. **James, D. J.**, Adventitious root formation in vitro in apple root stocks *(Malus pumila)*. I. Factors affecting the length of the auxin sensitive phase in M9, *Physiol. Plant.*, 57, 149, 1983.

146. **Jelaska, S.**, Embryogenesis and organogenesis in pumpkin explants, *Physiol. Plant.*, 31, 257, 1974.

147. **Jha, S., Mitra, G. C., and Sen, S.**, In vitro regeneration from bulb explants of Indian squill *Urginea indica, Plant Cell Tissue Organ Culture*, 3, 91, 1984.

148. **Johri, B. M. and Bhojwani, S. S.**, Growth responses of mature endosperm in cultures, *Nature (London)*, 208, 1345, 1965.

149. **Johri, B. M. and Nag, K. K.**, Endosperm of *Taxillus vestitus*, a system to study the effects of cytokinins in vitro in shoot-bud formation, *Curr. Sci.*, 39, 177, 1970.

150. **Jones, O. P.**, Effect of phloridzin and phloroglucinol on apple shoots, *Nature (London)*, 262, 392, 1976.

151. **Jones, O. P. and Hopwood, M. E.**, The successful propagation in vitro of two root stocks of *Prunus:* the root stock 'Pixy' *(P. institia)* and cherry root stock F 12/1 *(P. avium), J. Hortic. Sci.*, 54, 63, 1979.

152. **Kadkade, P. and Seibert, M.**, Phytochrome-regulated organogenesis in lettuce tissue culture, *Nature (London)*, 270, 49, 1977.

153. **Kadkade, P. G., Wetherebee, P., Jopson, H., and Botticelli, C.**, Abstr. 4th Int. Congr. Plant Tissue Cell Culture, Calgary, Canada, 1978, 29.

154. **Kaiser, W. J. and Teemba, L. R.**, Use of tissue culture and thermotherapy to free east African cassava cultivars of African Cassava mosaic and Cassava brown streak diseases, *Plant Dis. Rep.*, 63, 780, 1979.

155. **Kao, K. N. and Michayluk, M. R.**, Embryoid formation in alfalfa cell suspension from different plants, *In Vitro*, 17, 645, 1981.

156. **Kamada, H. and Harada, H.**, Studies on organogenesis in carrot tissue culture. II. Effects of amino acids and inorganic nitrogenous compounds on somatic embryogenesis, *Z. Pflanzenphysiol.*, 91, 453, 1979.

157. **Kamada, H. and Harada, H.**, Changes in endogenous levels and the effects of abscisic acid during somatic embryogenesis of *Daucus carota, Plant Cell Physiol.*, 22, 1423, 1981.

158. **Kamada, H. and Harada, H.**, Studies on nitrogen metabolism during somatic embryogenesis in carrot. I. Utilization of α-alanine as a nitrogen source, *Plant Sci. Lett.*, 33, 7, 1984.

159. **Kartha, K. K., Gamborg, O. L., and Constabel, F.**, In vitro plant formation from stem explants of rape *(Brassica napus* cv. *zephyr), Physiol, Plant.*, 31, 217, 1974.

160. **Kassanis, B.**, The use of tissue culture to produce virus-free clones from infected potato varieties, *Ann. Appl. Biol.*, 45, 422, 1957.

161. **Kato, H. and Takeuchi, M.**, Embryogenesis from the epidermal cells of carrot hypocotyl, *Sci. Rep. Coll. Gen. Educ. Univ. Tokyo*, 16, 245, 1966.

162. **Kavathekar, A. K., Ganapathy, P. S., and Johri, B. M.**, In vitro responses of embryoids of *Eschscholzia californica, Biol. Plant.*, 20, 98, 1978.

163. **Kerbauy, G. B.**, In vitro flowering of *Oncidium varicosum* mericlones (Orchidaceae), *Plant Sci. Lett.*, 35, 73, 1984.

164. **Kerns, H. R., Barwale, U. B., Meyer, M. M., and Widholm, J. M.**, Correlation of cotyledonary node shoot proliferation and somatic embryoid development in suspension cultures of soybean *(Glycine max), Plant Cell Rep.*, 5, 140, 1986.

165. **Kessel, R. H. J. and Carr, A. H.,** The effects of dissolved oxygen concentration on growth and differentiation of carrot *(Daucus carota)* tissue, *J. Exp. Bot.*, 23, 996, 1972.

166. **Kikuta, Y. and Okazawa, Y.,** Control of root and shoot bud formation from potato tuber tissue cultured in vitro, *Physiol. Plant.*, 61, 8, 1984.

167. **Kochba, J. and Spigel-Roy, P.,** Effect of culture media on embryoid formation from ovular callus of 'Shamouti' orange *Citrus sinensis*, *Z. Pflanzenzuecht.*, 69, 156, 1973.

168. **Kochba, J. and Spigel-Roy, P.,** The effects of auxins, cytokinins and inhibitors on embryogenesis in habituated ovular callus of 'Shamouti' orange *Citrus sinensis*, *Z. Pflanzenphysiol.*, 81, 283, 1977.

169. **Kochba, J. and Spigel-Roy, P.,** Cell and tissue culture for breeding and developmental studies of *Citrus*, *Hortic. Sci.*, 12, 110, 1977.

170. **Kochba, J., Spigel-Roy, P., and Safran, H.,** Adventive plants from ovules and nucelli in *Citrus*, *Planta*, 106, 237, 1972.

171. **Kochba, J., Spigel-Roy, P., Neumann, H., and Saad, S.,** Stimulation of embryogenesis in citrus ovular callus by ABA, ethepon, CCC, and alar and its suppression by GA, *Z. Pflanzenphysiol.*, 89, 427, 1978.

172. **Kochba, J., Button, J., Spigel-Roy, P., Bornman, C. H., and Kocba, M.,** Stimulation of rooting of citrus embryos by gibberellic acid and adenine sulphate, *Ann. Bot.*, 38, 795, 1974.

173. **Kochhar, T. S., Bhalla, P. R., and Sabharwal, P. S.,** Effect of tobacco smoke components on organogenesis in plant tissues, *Plant Cell Physiol.*, 12, 602, 1971.

174. **Kohlenbach, H. W.,** Über organisierte Bildungen aus *Macleya cordata* Kallus, *Planta*, 64, 37, 1965.

175. **Konar, R. N. and Konar, A.,** Plantlet and flower formation in callus cultures from *Phlox drummondii*, *Phytomorphology*, 16, 379, 1966.

176. **Konar, R. N. and Nataraja, K.,** Experimental studies in *Ranunculus sceleratus*. Development of embryos from stem epidermis, *Phytomorphology*, 15, 132, 1965.

177. **Konar, R. N. and Nataraja, K.,** Morphogenesis of isolated floral buds of *Ranunculus sceleratus* in vitro, *Acta Bot. Neerland*, 18, 680, 1969.

178. **Konar, R. N., Thomas, E., and Street, H. E.,** Origin and structure of embryoids arising from epidermal cells of the stem of *Ranunculus sceleratus*, *J. Cell Sci.*, 11, 77, 1972.

179. **Kononowicz, A. K. and Janick, J.,** The influence of carbon source on the growth and development of asexual embryos of *Theobroma caco*, *Physiol. Plant.*, 61, 155, 1984.

180. **Kononowicz, H., Kononowicz, A. K., and Janick, J.,** Asexual embryogenesis via callus of *Theobroma cacao*, *Z. Pflanzenphysiol.*, 113, 347, 1984.

181. **Krikorian, A. D. and Kann, R. P.,** Plantlet production from morphogenetically competent cell suspension of daylily, *Ann. Bot.*, 47, 679, 1981.

182. **Kulescha, Z.,** Etude de la degradation l'absorption et la fixation de l'acide 3-indoyl acetique par le tissu de topinambour, in *La Culture des Tissus et des Cellules des Vegetaux*, Gautheret, R. J., Ed., Masson, Paris, 1977, 59.

183. **Kumar, P. K., Raj, C. R., Chandramohan, M., and Iyer, R. D.,** Induction and maintenance of friable callus from cellular endosperm of *Cocos nucifera*, *Plant Sci.*, 40, 203, 1985.

184. **Lakshmi Sita, G., Raghava Ram, N. V., and Vaidyanathan, C. S.,** Triploid plants from endosperm cultures of sandalwood by experimental embryogenesis, *Plant Sci. Lett.*, 20, 63, 1980.

185. **Lazar, M. D., Collins, G. B., and Vida, W. E.,** Genetic and environmental effects on the growth and differentiation of wheat somatic cell cultures, *J. Hered.*, 74, 353, 1983.

186. **Lazzeri, P. A. and Dunwell, J. M.,** Establishment of isolated root cultures of *Brassica* species and regeneration from cultured root segments of *Brassica oleracea* var. *italica*, *Ann. Bot.*, 54, 351, 1984.

187. **Leu, L. S.,** Freeing sugarcane from mosaic virus by apical meristem culture and tissue culture, *Rep. Taiwan Sugar Exp. Stn.*, 57, 57, 1972.

188. **Leung, D. W. M. and Thorpe, T. A.,** Occurrence of α-amylase in tobacco callus: some properties and relationships with callus induction and shoot primordium formation, *J. Plant Physiol.*, 117, 363, 1985.

189. **Levine, M.,** The growth of normal plant tissue in vitro as affected by chemical carcinogens and plant growth substances. III. The culture of sunflower and tobacco stem segments, *Bull. Torrey Bot. Club*, 77, 110, 1950.

190. **Ling, D. H., Chen, W. Y., Chen, M. F., and Ma, Z. R.,** Direct development of plantlets from immature panicles of rice in vitro, *Plant Cell Rep.*, 2, 172, 1983.

191. **Litz, R. E. and Conover, R. A.,** Somatic embryogenesis in cell cultures of *Carica stipulata*, *Hortic. Sci.*, 15, 733, 1980.

192. **Litz, R. E. and Conover, R. A.,** In vitro somatic embryogenesis and plantlet regeneration from *Carica papaya* ovular callus, *Plant Sci. Lett.*, 26, 153, 1982.

193. **Liu, J. R. and Cantliffe, D. J.,** Somatic embryogenesis and plant regeneration in tissue cultures of sweet potato *(Ipomea batatas)*, *Plant Cell Rep.*, 3, 112, 1984.

194. **Lorz, H., Potrykus, I., and Thomas, E.,** Somatic embryogenesis from tobacco protoplasts, *Naturwissenschaften*, 64, 439, 1977.

195. **Lowe, K., Taylor, D. B., Ryan, P., and Paterson, K. E.**, Plant regeneration via organogenesis and embryogenesis in the maize inbred line B 73, *Plant Sci.*, 41, 125, 1984.

196. **Lu, C. and Vasil, I. K.**, Somatic embryogenesis and plantlet regeneration from leaf tissues of *Panicum maximum*, *Theor. Appl. Genet.*, 59, 275, 1981.

197. **Lu, C., Vasil, I. K., and Ozias-Akins, P.**, Somatic embryogenesis in *Zea mays*, *Theor. Appl. Genet.*, 62, 109, 1982.

198. **Lu, C-Y., Chandler, S. F., and Vasil, I. K.**, Somatic embryogenesis and plant regeneration from cultured immature embryos of rye *(Secale cereale)*, *J. Plant Physiol.*, 115, 237, 1985.

199. **Lupotto, E.**, Propagation of an embryogenic culture of *Medicago sativa*, *Z. Pflanzenphysiol.*, 111, 95, 1983.

200. **Lupotto, E.**, The use of single somatic embryo culture in propagating and regenerating lucerne *Medicago sativa*, *Ann. Bot.*, 57, 19, 1986.

201. **Maan, A. C., Van der Linde, P. C. G., Harkes, P. A. A., and Libbenga, K. R. L.**, Correlation between the presence of membrane bound auxin binding and root regeneration in cultured tobacco cells, *Planta*, 164, 376, 1985.

202. **Magnusson, I. and Bornman, C. H.**, Anatomical observations on somatic embryogenesis from scuteller tissues of immature zygotic embryos of *Triticum aestivum*, *Physiol. Plant.*, 63, 137, 1985.

203. **Maheswaran, G. and Williams, E. G.**, Direct somatic embryoid formation on immature embryos of *Trifolium repens, T. pratense* and *Medicago sativa* and rapid clonal propagation of *T. repens*, *Ann. Bot.*, 54, 201, 1984.

204. **Maheswaran, G. and Williams, E. G.**, Origin and development of somatic embryo formed directly on immature embryos of *Trifolium repens* in vitro, *Ann. Bot.*, 56, 619, 1985.

205. **Maheshwari, P. and Baldev, B.**, In vitro induction of adventive buds from embryos of *Cuscuta reflexa*, in, *Plant Embryology — A Symposium*, CSIR, New Delhi, 1961, 129.

206. **Maraffa, S. B., Sharp, W. R., Tayama, H. K., and Fretz, T. A.**, Apparent asexual embryogenesis in cultured leaf sections of Hoya, *Z. Pflanzenphysiol.*, 102, 45, 1981.

207. **Margara, J.**, *C. R. Acad. Sci. Ser. D*, 284, 1883, 1977.

208. **Mayer, L.**, Wachstum und organbildung an in vitro kulturvierten segmenten von *Pelargonium zonale* und *Cyclamen Persicum*, *Planta*, 47, 401, 1956.

209. **McComb, J. A.**, Clonal propagation of woody plants species with special reference to apples, *Proc. Int. Plant Prop. Soc.*, 28, 413, 1978.

210. **McWilliams, A., Smith, S. M., and Street, H. E.**, The origin and development of embryoids in suspension cultures of carrot, *Ann. Bot.*, 38, 243, 1974.

211. **Mehra, P. N. and Jaidka, K.**, Experimental induction of embryogenesis in pear, *Phytomorphology*, 35, 1, 1985.

212. **Mehra, P. N. and Sachdeva, S.**, Embryogenesis in apple in vitro, *Phytomorphology*, 34, 26, 1984.

213. **Miller, C. O., Skoog, F., Saltza, M., and Strong, F. M.**, Kinetin a cell division factor from deoxyribonucleic acid, *J. Am. Chem. Soc.*, 77, 1392, 1955.

214. **Mitra, J., Mapes, M. O., and Steward, F. C.**, Growth and organized development of cultured cells. IV The behaviour of the nucleus, *Am. J. Bot.*, 47, 357, 1960.

215. **Montague, M. S., Armstrong, T. A., and Jaworski, E. G.**, Polyamine metabolism in embryogenic cells of *Daucus carota*. II. Changes in arginine decarboxylase activity, *Plant Physiol.*, 63, 341, 1979.

216. **Montague, M. J., Koppenbrink, J. W., and Jaworski, E. G.**, Polyamine metabolism in embryogenic cells of *Daucus carota*. I. Changes in intracellular content and rate of synthesis, *Plant Physiol.*, 62, 430, 1978.

217. **Morel, G.**, Producing virus-free *Cymbidium*, *Bull. Am. Orchid Soc.*, 29, 495, 1960.

218. **Morel, G.**, Morphogenesis of stem apical meristem cultivated in vitro: application to clonal propagation, *Phytomorphology*, 22, 265, 1972.

219. **Morel, G. and Martin, C.**, Guerison de pommes de terre atteintics de maladies a virus, *C.R. Acad. Agric. Fr.*, 41, 471, 1952.

220. **Mori, K.**, Production of virus-free plants by means of meristem culture, *Jap. Agric. Res. Q.*, 6, 1, 1971.

221. **Mouras, A., and Lutz, A.**, Induction, repression et conservation des proprietes embryogenetiques des cultures de tissue de carotte sauvage, *Bull. Soc. Bot. Fr.*, 127, 93, 1980.

222. **Murashige, T.**, Analysis of the inhibition of organ formation in tobacco tissue culture by gibberellin, *Physiol. Plant.*, 17, 636, 1964.

223. **Murashige, T.**, Effects of stem elongation retardants and gibberellin on callus growth and organ formation in tobacco tissue culture, *Physiol. Plant.*, 18, 665, 1965.

224. **Murashige, T.**, Plant propagation through tissue cultures, Ann. Rev. *Plant Physiol.*, 25, 135, 1974.

225. **Murashige, T. and Nakano, R.**, Chromosome complement as a determinant of the morphogenic potential of tobacco cells, *Am. J. Bot.*, 54, 963, 1967.

226. **Murashige, T., Serpa, M., and Jones, J. B.**, Clonal multiplication of gerbera through tissue culture, *Hortic. Sci.*, 9, 175, 1974.

227. **Murashige, T., Shabde, N. N., Hasegawa, P. M., Takatoi, F. H., and Jones, J. B.,** Propagation of *Asparagus* through shoot apex culture. I. Nutrient medium for formation of plantlets, *J. Am. Soc. Hortic. Sci.,* 97, 158, 1972.

228. **Murashige, T. and Skoog, F.,** A revised medium for rapid growth and bioassays with tobacco tissue cultures, *Physiol. Plant.,* 15, 473, 1962.

229. **Murray, B. E., Handyside, R. J., and Keller, W. A.,** In vitro regeneration of shoots on stem explants of haploid and diploid flax *(Linum usitatissimum), Can. J. Genet. Cytol.,* 19, 177, 1978.

230. **Nag, K. K. and Johri, B. M.,** Morphogenetic studies on endosperm of some parasitic angiosperms, *Phytomorphology,* 21, 202, 1971.

231. **Nagarajan, P., Mckenzie, J. S., and Walton, P. D.,** Embryogenesis and plant regeneration of *Medicago* spp. in tissue culture, *Plant Cell Rep.,* 5, 77, 1986.

232. **Nair, S., Shirgurkar, M. V., and Mascerenhas, A. F.,** Studies on endosperm culture of *Annona squamosa, Plant Cell Rep.,* 5, 132, 1986.

233. **Narasimhulu, S. B. and Reddy, G. M.,** In vitro flowering and pod formation from cotyledons of groundnut *(Arachis hypogea), Theor. Appl. Genet.,* 69, 87, 1984.

234. **Narayanaswami, S.,** Regeneration of plants from tissue cultures, in *Applied and Fundamental Aspects of Plant Cell Tissue and Organ Culture,* Reinert, J. and Bajaj, Y. P. S., Eds., Springer-Verlag, Berlin, 1977, 179.

235. **Negrutiu, I., Jacobs, M., and Cachita, D.,** Some factors controlling in vitro morphogenesis of *Arabidopsis thaliana, Z. Pflanzenphysiol.,* 86, 113, 1978.

236. **Newell, C. A. and Luu, H. T.,** Protoplast culture and plant regeneration in *Glycine canescens, Plant Cell Tissue Organ Culture,* 4, 145, 1985.

237. **Nitsch, C.,** Induction in vitro de la floraison chez une plante de Jours courts: *Plumbago indica, Ann. Sci. Nat. (Bot.),* 9, 1, 1968.

238. **Nitsch, J. P. and Nitsch, C.,** Neoformation de boutons floraux sur cultures in vitro de feuilles det de racines de *Chicorium intybus.* Existence de uretat vernalise en l' absence de bourgeons, *Bull. Soc. Bot. Fr.,* 111, 299, 1964.

239. **Nobercourt, P.,** Sur la perennite et l'augmentation de volume des culture de tissus vegetaux, *C.R. Soc. Biol.,* 130, 1270, 1939.

240. **Nobercourt, P.,** Sur les radicelles naissant des cultures de tissue vegetaux, *C.R. Soc. Biol.,* 130, 1271, 1939.

241. **Noma, M., Huber, J., Ernst, D., and Pharis, R. P.,** Quantitation of gibberellins and the metabolism of ^{3}H gibberellin A_1 during embryogenesis in carrot and anise cell cultures, *Planta,* 155, 369, 1982.

242. **Noma, M., Koike, N., Sano, M., and Kawashima, N.,** Endogenous indole-3-acetic acid in the stem of tobacco in relation to flower neoformation as measured by mass spectroscopic assay, *Plant Physiol.,* 75, 257, 1984.

243. **Nomura, K. and Komamine, A.,** Identification and isolation single cells that produce somatic embryos at a high frequency in a carrot suspension culture, *Plant Physiol.,* 79, 988, 1985.

244. **Nowicki, M. E. and O'Rourke, E. N.,** On the elimination of amaryllus mosaic virus from Leopoldi amaryllus hybrids. I. Shoot apex culture, *Plant Life,* 30, 108, 1974.

245. **Nwankwo, B. A. and Krikorian, A. D.,** Morphogenetic potential of embryo and seedling-derived callus of *Elaeis quineensis, Ann. Bot.,* 51, 65, 1983.

246. **Osborne, D. J.,** Control of cell shape and cell size by the dual regulation of auxin and ethylene, in *Perspectives in Experimental Biology,* Sunderland, N., Ed., Pergamon Press, Oxford, 1976, 89.

247. **Ozias-Akins, P. and Vasil, I. K.,** Plant regeneration from cultured immature embryos and inflorescences of *Triticum aestivum* (wheat). Evidence for somatic embryogenesis, *Protoplasma,* 110, 95, 1982.

248. **Ozias-Akins, P. and Vasil, I. K.,** Proliferation of and plant regeneration from epiblast of *Tritium aestivum* (wheat, Gramineae) embryos, *Am. J. Bot.,* 70, 1092, 1983.

249. **Paludan, N.,** Etablering aus virus-free meristem kulture und havebrugsplanter, *Tidsskr. Planteavl,* 75, 387, 1971.

250. **Partanen, C. R.,** Quantitative chromosomal changes and differentiation in plants, in *Developmental Cytology,* Rudnick, D., Ed., Ronald Press, New York, 1959, 21.

251. **Perez-Brumudez, P., Brisa, M. C., Cornejo, M. J., and Segura, J.,** In vitro morphogenesis from excised leaf explants of *Digitalis obscura, Plant Cell Rep.,* 3, 8, 1984.

252. **Perez-Bermudez, P., Cornejo, M. J., and Segura, J.,** A morphogenetic role for ethylene in hypocotyl cultures of *Digitalis obscura, Plant Cell Rep.,* 4, 188, 1985.

253. **Peterson, R. L.,** The initiation and development of root buds, in *The Development and Function of Roots,* Torrey, J. G. and Clarkson, D. T., Eds., Academic Press, New York, 1975, 125.

254. **Pfaff, W. and Schopfer, P.,** Hormones are no causal links in phytochrome-mediated adventitious root formation in mustard seedlings *(Sinapsis alba), Planta,* 150, 321, 1980.

255. **Pierik, R. L. M.,** Regeneration, vernalization and flowering in *Lunaria annua* in vivo and in vitro, *Meded. Landbouwhogesch Wageningen,* 67, 60, 1967.

256. **Pierik, R. L. M. and Steegmans, H. H. M.**, Analysis of adventitious root formation in isolated stem explants of rhododendron, *Sci. Hortic.*, 3, 1, 1975.

257. **Pillai, S. K.**, Alternating light and dark periods in the differentiation of Geranium callus, in *Seminar on Plant Morphogenesis*, University of Delhi, India, 1968, 66.

258. **Price, H. J. and Smith, R. H.**, Somatic embryogenesis in suspension cultures of *Gossypium klotzschainum*, *Planta*, 145, 305, 1979.

259. **Pua, E.-C., Ragolsky, E., and Thorpe, T. A.**, Retention of shoot regeneration capacity of tobacco callus by Na_2SO_4, *Plant Cell Rep.*, 4, 225, 1985.

260. **Quak, F.**, Meristem culture and virus-free plants, in *Applied and Fundamental Aspects of Plant Cell Tissue and Organ Culture*, Reinert, J. and Bajaj, Y. P. S., Eds., Springer-Verlag, Berlin, 1977, 598.

261. **Radojevic, L.**, Tissue culture of maize *Zea mays* 'cudu'. I. Somatic embryogenesis in the callus tissue, *J. Plant Physiol.*, 119, 435, 1985.

262. **Radojevic, L. J., Kavoor, J., and Zylberg, L.**, Etude anatomique et histochemique des cals embryogenes du *Corylus avellana* et du *Paulownia tomentosa*, *Rev. Cytol. Biol. Veg.*, 2, 155, 1979.

263. **Raghavan, V.**, *Experimental Embryogenesis in Vascular Plants*, Academic Press, London, 1976.

264. **Raju, M. Y. S. and Mann, H. E.**, Regenerative studies in the detached leaves of *Echeveria elegans*. Anatomy and regeneration of leaves in sterile culture, *Can. J. Bot.*, 48, 1887, 1970.

265. **Rangan, T. S., Murashige, T., and Bitters, W. P.**, In vitro Initiation of nucellar embryos in monoembryonate citrus, *Hortic. Sci.*, 3, 226, 1968.

266. **Rangaswamy, N. S.**, Experimental studies on female reproductive structures of *Citrus microcarpa*, *Phytomorphology*, 11, 109, 1961.

267. **Rao, I. U., Rao, I. V., and Narang, V.**, Somatic embryogenesis and regeneration of plants in the bamboo *Dendrocalamus strictus*, *Plant Cell Rep.*, 4, 191, 1985.

268. **Rao, P. S., and Narayanaswami, S.**, Morphogenetic investigations in callus cultures of *Tylophora indica*, *Physiol. Plant.*, 27, 271, 1972.

269. **Rapela, M. A.**, Organogenesis and somatic embryogenesis in tissue cultures of Argentina maize *Zea mays*, *J. Plant Physiol.*, 121, 119, 1985.

270. **Rashid, A. and Reinert, J.**, 1980. Selection of embryogenic pollen from cold treated buds of *Nicotiana tabacum* var. Badischer Burley and their development into embryos in cultures, *Protoplasma*, 105, 161, 1980.

271. **Rashid, A. and Reinert, J.**, High frequency embryogenesis in *ab initio* pollen cultures of *Nicotiana*, *Naturwissenschaften*, 68, 378, 1981.

272. **Rashid, A. and Street, H. E.**, Growth and embryogenic potential and stability of haploid cell culture of *Atropa belladonna*, *Plant Sci. Lett.*, 2, 89, 1974.

273. **Reisch, B. and Bingham, E. T.**, The genetic control of bud formation from callus cultures of diploid alfalfa, *Plant Sci. Lett.*, 20, 71, 1980.

274. **Reinert, J.**, Morphogenese und ihre kontrolle an Gewebekulturen aus karotten, *Naturwissenschaften*, 45, 344, 1958.

275. **Reinert, J.**, Untersuchungen über die Morphogenese an Gewebekulturen, *Ber. Dtsch. Bot. Ges.*, 71, 15, 1958.

276. **Reinert, J.**, Über die Kontrolle der Morphogense und die Induktion von Adventivembryonen an Gewebekulturen aus Krotten, *Planta*, 53, 318, 1959.

277. **Reinert, J.**, Differenzirung und Bildung von Adventivembryonen aus einzelnen Zellen von Karottengewebe, *Abstr. 10th Int. Bot. Congr.*, Edinburgh, 1964, 209.

278. **Reinert, J., Backs-Husemann, D., and Zerman, H.**, Determination of embryo and root formation in tissue cultures from *Daucus carota*, in *Les Cultures de Tissus de Plantes (Colloq. Int. C.N.R.S.)*, No. 193, 1971, 261.

279. **Reinert, J., Bajaj, Y. P. S., and Zbell, B.**, Aspects of organization — organogenesis embryogenesis, cytodifferentiation, in *Plant Cell, Tissue and Organ Culture*, Street, H. E., Ed., Blackwell, Oxford, 1977, 389.

280. **Reinert, J., Tazawa, M., and Somenoff, S.**, Nitrogen compounds as factors of embryogenesis in vitro, *Nature (London)*, 216, 1215, 1967.

281. **Reuther, G. and Bhandari, N. N.**, Organogenesis and histogenesis of adventitious organs induced on leaf blade segments of Begonia-Elatior hybrids *(Begonia × hiemotis)* in tissue culture, *Gartenbauwissenschaft*, 46, 241, 1981.

282. **Reynold, J. F. and Murashige, T.**, Asexual embryogenesis in callus cultures of palms, *In Vitro*, 15, 383, 1979.

283. **Riviere, S.**, Culture in vitro du meristeme apical du *Lilium candidum* L. (Liliacees), apres ablation totale des primordiums foliarixes, *C.R. Acad. Sci. Ser. D*, 276, 1989, 1973.

284. **Roest, S. and Bokelmann, G. S.**, Vegetative propagation of *Chrysanthemum cinerariefolium* in vitro, *Sci. Hortic.*, 1, 120, 1973.

285. **Ronchi, V. N., Caligo, M. A., Nozzolini, M., and Lucearini, G.,** Stimulation of carrot somatic embryogenesis by proline and serine, *Plant Cell Rep.,* 3, 210, 1984.

286. **Ross, M. K. and Thorpe, T. A.,** Physiological gradients and shoot initiation in tobacco callus culture, *Plant Cell Physiol.,* 14, 473, 1973.

287. **Ross, M. K., Thorpe, T. A., and Costerton, J. W.,** Ultrastructural aspects of shoot initiation in tobacco callus cultures, *Am. J. Bot.,* 60, 788, 1973.

288. **Rucker, W.,** Combined influence of indoleacetic acid, gibberellic acid and benzylaminopurine on callus and organ differentiation in *Digitalis purpurea* leaf explants, *Z. Pflanzenphysiol.,* 107, 141, 1982.

289. **Rucker, W. and Radola, B.,** Isoelectric patterns of peroxidase isoenzymes from tobacco tissue cultures, *Planta,* 99, 192, 1971.

290. **Sabharwal, P. S.,** In vitro culture of ovules, nucelli and embryos of *Citrus reticulata,* in *Plant Tissue and Organ Culture — A Symposium,* Maheshwari, P. and Rangaswaney, N. S., Eds., Int. Soc. of Plant Morphologists, Delhi, India, 1963, 255.

291. **Sastri, R. L. N.,** Morphogenesis in plant tissue cultures, in *Plant Tissue and Organ Culture — a Symposium,* Maheshwari, P. and Rangaswaney, N. S. Eds., Int. Soc. of Plant Morphologists, Delhi, India, 1963, 105.

292. **Schiavone, F. M., and Cooke, T. J.,** A genometric analysis of somatic embryo formation in carrot cell cultures, *Can. J. Bot.,* 63, 1573, 1985.

293. **Schultz, P. and Jensen, W. A.,** *Capsella* embryogenesis, the early embryo, *J. Ultra. struct. Res.,* 22, 376, 1968.

294. **Seabrook, J. E. A., Cumming, B. G., and Dionne, L. A.,** The in vitro induction of adventitious shoot and root apices on Narcissus (daffodil and narcissus) cultivar tissue, *Can. J. Bot.,* 54, 814, 1976.

295. **Sehgal, C. B. and Khurana, S.,** Morphogenesis and plant regeneration from cultured endosperm of *Emblica officinalis, Plant Cell Rep.,* 4, 263, 1985.

296. **Seibert, M., Wetherbee, P. J., and Job, D. B.,** The effects of light intensity and spectral quality on growth and shoot initiation in tobacco callus, *Plant Physiol.,* 56, 130, 1975.

297. **Sekiya, J. and Yamada, Y.,** *Bull. Inst. Chem. Res. Kyoto Univ.,* 52, 246, 1974.

298. **Sengupta, C. and Raghavan, V.,** Somatic embryogenesis in carrot cell suspension. II. Synthesis of ribosomal RNA and polyA + RNA, *J. Exp. Bot.,* 31, 259, 1980.

299. **Shabde, M. and Murashige, T.,** Hormonal requirement of excised *Dianthus caryophyllus* shoot apical meristem in vitro, *Am. J. Bot.,* 64, 443, 1977.

300. **Sharp, W. R., Sondhall, M. R., Caldas, L. S., and Maraffa, S. B.,** The physiology of in vitro asexual embryogenesis, *Hortic. Rev.,* 2, 268, 1980.

301. **Shepard, J. F., Bidney, D., and Shahin, E.,** Potato protoplasts in crop improvement, *Science,* 208, 17, 1980.

302. **Simonsen, J. and Hildebrandt, A. C.,** In vitro growth and differentiation of *Gladiolus* plants from callus cultures, *Can. J. Bot.* 49, 1817, 1971.

303. **Sinnott, E. W.,** The cell organ relationship in plant organization, *Growth Symp.,* 3, 77, 1939.

304. **Siriwardana, S. and Nabors, M. W.,** Tryptophan enhancement of somatic embryogenesis in rice, *Plant Physiol.,* 73, 142, 1983.

305. **Skokut, T. A., Manchester, J., and Schaefer, J.,** Regeneration in alfalfa tissue culture. Stimulation of somatic embryo produced by amino acids and N-15 NMR determination of nitrogen utilization, *Plant Physiol.,* 79, 579, 1985.

306. **Skoog, F.,** Growth and organ formation in tobacco tissue culture, *Am. J. Bot.,* 31, 19, 1944.

307. **Skoog, F.,** Aspects of growth factor interactions in morphogenesis in tobacco tissue cultures, in *Les Cultures de Tissus de Plantes (Colloq. Int. C.N.R.S.),* No. 193, 1971, 115.

308. **Skoog, F. and Miller, C. O.,** Chemical regulation of growth organ formation in plant tissue cultures in vitro, *Symp. Soc. Exp. Biol.,* 11, 118, 1957.

309. **Skoog, F. and Tsui, C.,** Chemical control of growth and bud formation in tobacco stem segments and callus cultured in vitro, *Am. J. Bot.,* 35, 782, 1948.

310. **Skoog, F. and Tsui, C.,** Growth substances and the formation of buds in plant tissues, in *Plant Growth Substances,* Skoog, F., Ed., University of Wisconsin Press, Madison, 1951, 263.

311. **Smith, R. H.,** In Vitro Development of the Isolated Shoot, Apical Meristem of Angiosperm, Ph.D. thesis, University of California, Riverside, 1970.

312. **Smith, R. H. and Murashige, T.,** In vitro development of isolated shoot apical meristems of angiosperms, *Am. J. Bot.,* 57, 562, 1970.

313. **Smith, R. H. and Murashige, T.,** Primordial leaf and phytohormone effects on excised shoot apical meristem of *Coleus blumei, Am. J. Bot.,* 69, 1334, 1982.

314. **Smith, S. M.,** Embryogenesis in Tissue Cultures of Domestic Carrot *Daucus carota,* Ph.D. thesis, University of Leicester, England, 1973.

315. **Smith, S. M. and Street, H. E.,** The decline of embryogenic potential as callus and suspension cultures of carrot *(Daucus carota)* are serially cultured, *Ann. Bot.,* 38, 233, 1974.

316. **Sondahl, M. R. and Sharp, W. R.,** High frequency induction of somatic embryos in cultured leaf explants of *Coffea arabica, Z. Pflanzenphysiol.,* 81, 395, 1977.

317. **Sriskandarajah, Müllins, M. G., and Nair, Y.,** Induction of adventitious rooting in vitro in difficult to propagate cultivars of apple, *Plant Sci. Lett.,* 24, 1, 1982.

318. **Srinivasan, C. and Müllins, M. G.,** High frequency somatic embryo production from unfertilized ovules of grapes, *Sci. Hortic.,* 13, 245, 1980.

319. **Srivastava, P. S.,** In vitro induction of triploid roots and shoots from mature endosperm of *Jatropha panduraefolia, Z. Pflanzenphysiol.,* 66, 93, 1971.

320. **Srivastava, P. S.,** Formation of triploid plantlets in endosperm cultures of *Putranjiva roxburghii, Z. Pflanzenphysiol.,* 69, 270, 1973.

321. **Stamp, J. A. and Henshaw, G. G.,** Somatic embryogenesis in Cassava, *Z. Pflanzenphysiol.,* 105, 183, 1982.

322. **Starisky, G.,** Embryoid formation in callus culture of coffee, *Acta Bot. Neerl.,* 19, 509, 1970.

323. **Stellwagen, R. H. and Cole, R. D.,** Chromosomal proteins, *Ann. Rev. Biochem.,* 38, 951, 1969.

324. **Steward, F. C.,** Carrots and coconuts: some investigations on growth, in *Plant Tissue and Organ Culture — A Symp.,* International Society of Plant Morphologists, Delhi, India, 1963.

325. **Steward, F. C., Kent, A. E., and Mapes, M. O.,** The culture of free plant cells and its significance for embryology and morphogenesis, in *Current Topics in Developmental Biology,* Academic Press, New York, 113, 1966.

326. **Steward, F. C., Mapes, M. O., Kent, A. E., and Holsten, R. D.,** Growth and development of cultured plant cells, *Science,* 143, 20, 1964.

327. **Steward, F. C., Mapes, M. O., and Mears, K.,** Growth and organized development of cultured cells. II. Organization in cultures grown from freely suspended cells, *Am. J. Bot.,* 45, 705, 1958.

328. **Steward, F. C. and Shantz, E. M.,** The chemical regulation of growth: some substances and extracts which induced growth and morphogenesis, *Ann. Rev. Plant Physiol.,* 10, 379, 1959.

329. **Stichel, E.,** Gleichzeitige Induktion von Sprossen und wurzien an in vitro Kultiviraten Gewebestucken von *Cyclamen persicum, Planta,* 53, 293, 1959.

330. **Stone, O. M.,** The elimination of four viruses from carnation and sweet williams by meristem tip culture, *Ann. Appl. Biol.,* 62, 119, 1968.

331. **Stone, O. M.,** The elimination of viruses from *Narcissus tazella* V. Grand soter and rapid multiplication of virus free clones, *Ann. Appl. Biol.,* 73, 45, 1973.

332. **Stoutemeyer, V. T. and Britt, O. K.,** The behaviour of tissue culture from english and algerian ivy in different growth phases, *Am. J. Bot.,* 45, 805, 1965.

333. **Street, H. E.,** Embryogenesis and chemically induced organogenesis, in *Plant Cell and Tissue Culture: Principles and Applications,* Sharp, W. R., Larson, P. O., Paddock, E. F., and Raghavan, V., Eds., Ohio State University Press, Columbus, 1979, 123.

334. **Sung, Z. R. and Okimoto, R.,** Embryogenic proteins in somatic embryos of carrot, *Proc. Natl. Acad. Sci. U.S.A.,* 78, 3683, 1981.

335. **Suprasanna, P., Rao, K. V., and Reddy, G. M.,** Plant regeneration from glume calli of maize, *Theor. Appl. Genet.,* 72, 120, 1986.

336. **Suresh Kumar, A., Reddy, T. P., and Reddy, G. M.,** Plantlet regeneration from callus cultures of pigeon pea *Cajanus cajan, Plant Sci. Lett.,* 32, 271, 1983.

337. **Suresh Kumar, A., Reddy, T. P., and Reddy, G. M.,** Genetic analysis of certain in vitro and in vivo parameters in pigeon pea *Cajanus cajan, Theor. Appl. Genet.,* 70, 151, 1985.

338. **Syono, K.,** Changes in organ forming capacity of carrot root callus during subculture, *Plant Cell Physiol.,* 6, 403, 1965.

339. **Syono, K.,** Physiological and biochemical changes of carrot root callus during successive cultures, *Plant Cell Physiol.,* 6, 371, 1965.

340. **Takahashi, A., Sakuragi, Y., Kameda, H., and Ihizuka, K.,** Plant regeneration through somatic embryogenesis in barnyard grass *Echinochloa oryzicola, Plant Sci. Lett.,* 36, 161, 1984.

341. **Takayama, S. and Misawa, M.,** Factors affecting differentiation and growth in vitro and mass propagation scheme for *Begonia × hiemalis, Sci. Hortic.,* 16, 65, 1982.

342. **Takeda, J. and Senda, M.,** Relationship of the electrophysiological property of tobacco cultured cells to their regeneration ability, *Plant Cell Physiol.,* 25, 619, 1984.

343. **Takeno, K., Koshioka, M., Pharis, R. M., Rajasekaran, and Müllins, M. G.,** Endogenous gibberellin-like substances in somatic embryos of grape *(Vitis vinifera × Vitis rupestris)* in relation to embryogenesis and chilling requirement for subsequent development of mature embryos, *Plant Physiol.,* 73, 803, 1983.

344. **Tanimoto, S. and Harada, H.,** Hormonal control of morphogenesis in leaf explants of *Perilla frutescens, Ann. Bot.,* 45, 321, 1980.

345. **Tanimoto, S. and Harada, H.,** Chemical factors controlling floral bud formation of *Torenia* stem segments cultured in vitro. I. Effects of mineral nutrients and sugars, *Plant Cell Physiol.,* 22, 533, 1981.

346. **Tanimoto, S. and Harada, H.,** Chemical factors controlling floral bud formation of *Torenia* stem segments cultured in vitro. II. Effects of growth regulators, *Plant Cell Physiol.,* 22, 543, 1981.

347. **Tanimoto, S. and Harada, H.,** Effects of IAA zeatin, ammonium nitrate and sucrose on the intiation and development of floral buds in *Torenia* stem segments cultured in vitro, *Plant Cell Physiol.,* 22, 1553, 1981.

348. **Tanimoto, S. and Harada, H.,** Roles of auxin and cytokinin in organogenesis in *Torenia* stem segments cultured in vitro, *J. Plant Physiol.,* 115, 11, 1984.

349. **Tanimoto, S. and Harada, H.,** Involvement of calcium in adventitious bud initiation in *Torenia* stem segments, *Plant Cell Physiol.,* 27, 1, 1986.

350. **Tanimoto, S., Miyazaka, A., and Harada, H.,** Regulation by abscisic acid of in vitro flower formation in *Torenia* stem segments, *Plant Cell Physiol.,* 26, 675, 1985.

351. **Tanimoto, S., Satoh, S., Fuji, T., and Harada, H.,** Inhibition of cytokinin-induced adventitious bud initiation in *Torenia* stem segments by inhibitors of serine proteases, *Plant Cell Physiol.,* 25, 1161, 1984.

352. **Tazawa, M. and Reinert, J.,** Extracellular and intracellular chemical environments in relation to embryogenesis, *Protoplasma,* 68, 157, 1969.

353. **Tejovathi, G. and Anwar, S. Y.,** In vitro induction of capitula from cotyledons of *Carthamus tuictorius,* *Plant Sci. Lett.,* 36, 165, 1984.

354. **Templeton-Somers, K. M. and Collins, W. W.,** Heritability of regeneration in tissue cultures of sweet potato *(Ipomea batatas),* *Theor. Appl. Genet.,* 71, 835, 1986.

355. **Tewes, A., Wappler, A., Peschke, E., Garve, R., and Nover, L.,** Morphogenesis and embryogenesis in long term cultures of *Digitalis, Z. Pflanzenphysiol.,* 106, 311, 1982.

356. **Thomas, E. and Street, H. E.,** Organogenesis in cell suspension cultures of *Atropa belladonna, Ann. Bot.,* 34, 657, 1970.

357. **Thomas, E. and Street, H. E.,** Factors influencing morphogenesis in excised roots and suspension cultures of *Atropa belladonna, Ann. Bot.,* 36, 223, 1972.

358. **Thomas, E., Konar, R. N., and Street, H. E.,** The fine structure of embryogenic callus of *Ranunculus sceleratus, J. Cell Sci.,* 11, 95, 1972.

359. **Thomas, M. R. and Scott, K. J.,** Plant regeneration by somatic embryogenesis from callus initiated from immature embryos and immature inflorescences of *Hordeum vulgare, J. Plant Physiol.,* 121, 159, 1985.

360. **Thorpe, T. A.,** Organogenesis in vitro: structural, physiological and biochemical aspects, *Int. Rev. Cytol.,* Supp. 11A, 71, 1980.

361. **Thorpe, T. A.,** Morphogenesis and regeneration in tissue culture, in *Genetic Engineering Application to Agriculture,* Beltsville Symposia in Agricultural Research, Owens, L. D., Ed., Rowman & Allanheld, London, 1983, 285.

362. **Thorpe, T. A. and Beaudoin-Eagan, L. D.,** ^{14}C-metabolism during growth and shoot formation in tobacco callus cultures, *Z. Pflanzenphysiol.,* 113, 337, 1984.

363. **Thorpe, T. A. and Gespar, I.,** Changes in isoperoxidases during shoot formation in tobacco callus, *In Vitro,* 14, 522, 1978.

364. **Thorpe, R. A., Joy, R. W., and Leung, W. M.,** Starch turnover in shoot-forming tobacco callus, *Physiol. Plant.,* 66, 58, 1986.

365. **Thorpe, T. A. and Meier, D. D.,** Starch metabolism, respiration and shoot formation in tobacco callus cultures, *Physiol. Plant.,* 27, 365, 1972.

366. **Thorpe, T. A. and Meier, D. D.,** Effect of gibberellic acid and abscisic acid on shoot formation in tobacco callus cultures, *Physiol. Plant.,* 29, 121, 1973.

367. **Thorpe, T. A. and Meier, D. D.,** Starch metabolism in shoot forming tobacco callus, *J. Exp. Bot.,* 25, 288, 1974.

368. **Thorpe, T. A. and Meier, D. D.,** Effect of gibberellic acid on starch metabolism in tobacco cells cultured under shoot-forming conditions, *Phytomorphology,* 25, 238, 1975.

369. **Thorpe, T. A. and Murashige, T.,** Some histochemical changes underlying shoot initiation in tobacco callus cultures, *Can. J. Bot.,* 48, 277, 1970.

370. **Thorpe, T. A. and Murashige, T.,** Starch accumulation in shoot forming tobacco callus cultures, *Science,* 160, 421, 1968.

371. **Tisserat, B., Esau, E. B., and Murashige, T.,** Somatic embryogenesis in angiosperms, *Hortic. Rev.,* 1, 1, 1979.

372. **Torrey, J. G.,** Endogenous bud and root formation by isolated roots of *Convolvulus* grown in vitro, *Plant Physiol.,* 33, 258, 1958.

373. **Torrey, J. G.,** The initiation of organized development in plants, in *Advances in Morphogenesis,* Vol. 5, Abercrombie, M. and Brachet, J., Eds., Academic Press, New York, 1966, 39.

374. **Tran Thanh Van, K.,** Direct flower neoformation from superficial tissue of small explants of *Nicotiana tabacum, Planta,* 115, 87, 1973.

375. **Tran Thanh Van, K.,** In vitro control of de novo flower, bud, root and callus differentiation from excised epidermal tissues, *Nature (London),* 246, 44, 1973.

376. **Tran Thanh Van, K.,** Control of morphogenesis by inherent and exogenously applied factors in thin cell layers, *Int. Rev. Cytol.,* Suppl. 11A, 175, 1980.

377. **Tran Thanh Van, K.,** Control of morphogenesis in in vitro cultures, *Ann. Rev. Plant Physiol.,* 32, 291, 1981.

378. **Tulecke, W. and McGranhan, G.,** Somatic embryogenesis and plant regeneration from cotyledons of walnut, *Juglans regia, Plant Sci.,* 40, 57, 1985.

379. **Tulecke, W., Weinstein, L. H., Ratmer, A., and Laurencot, H. J.,** The biochemical composition of coconut water (coconut milk) as related to its use in plant tissue culture, *Contrib. Boyce Thompson Inst.,* 21, 115, 1961.

380. **Tyagi, A. K., Bharal, S., Rashid, A., and Maheshwari, N.,** Plant regeneration from tissue cultures initiated from immature inflorescences of a grass *Echinochloa colonum, Plant Cell Rep.,* 4, 115, 1985.

381. **Upadhyaya, S. and Chandra, N.,** Shoot and plantlet formation in organ and callus cultures of *Albizzia lebbek, Ann. Bot.,* 52, 421, 1983.

382. **Van Aartrij, K. J. and Blom-Barnhoorn, G. J.,** Growth regulator requirements for adventitious regeneration for *Lilium* bulb scale tissue in vitro in relation to duration of bulb storage and cultivar, *Sci. Hortic.,* 14, 261, 1981.

383. **Van Aartrij, K. J. and Blom-Barnhoorn, G. J.,** Adventitious bud formation from bulb-scale explants of *Lilium speciosum* in vitro. Interacting effects of NAA, TIBA, wounding and temperature, *J. Plant Physiol.,* 116, 409, 1984.

384. **Van Aartrij, K. J., Blom-Barnhoorn, G. J., and Bruinsma, J.,** Adventitious bud formation from bulb scale explants of *Lilium speciosum* in vitro. Effects of aminoethoxy vinyl glycine, 1-amino-cyclopropane-1-carboxylic acid and ethylene, *J. Plant Physiol.,* 117, 401, 1985.

385. **Van Aartrij, K. J., Blom-Barnhoorn, G. J., and Bruinsma, J.,** Adventitious bud formation from bulb scale explants of *Lilium speciosum* in vitro. Production of ethane and ethylene, *J. Plant Physiol.,* 117, 411, 1985.

386. **Vasil, V. and Hildebrandt, A. C.,** Differentiation of tobacco plants from single isolated cells in microculture, *Science,* 150, 889, 1965.

387. **Vasil, V., Lu, C-Y., and Vasil, I. K.,** Histology of somatic embryogenesis in cultured immature embryos of maize *(Zea mays), Protoplasma,* 127, 1, 1985.

388. **Vasil, V. and Vasil, I. K.,** Isolation and culture of cereal protoplasts. II. Embryogenesis and plantlet formation from protoplasts of *Pennisetum americanum, Theor. Appl. Genet.,* 56, 97, 1980.

389. **Vasil, V. and Vasil, I. K.,** Somatic embryogenesis and plant regeneration from culture of pearl millet *Pennisetum americanum* suspension, *Ann. Bot.,* 47, 649, 1981.

390. **Vij, S. P., Sood, A., and Plaha, K. K.,** Propagation of *Rhyncostylis retusa* (Orthidaceae) by direct organogenesis from leaf segment cultures, *Bot. Gaz.,* 145, 210, 1984.

391. **Vöchting, H.,** *Uber Organbildung in Pflanzenreich,* Vol. 1, Thiel Max Cohen, Bonn, 1878.

392. **Walker, K. A., Yu, P. C., Sato, S. J., and Jaworski, E. G.,** The hormonal control of organ formation in callus of *Medicago sativa* cultured in vitro, *Am. J. Bot.,* 65, 654, 1978.

393. **Walkey, D. G. A.,** In vitro methods for virus elimination in *Frontiers of Plant Tissue Culture,* Thorpe, T. A., Ed., University of Calgary Press, Canada, 1978, 245.

394. **Walkey, D. G. A. and Thompson, A.,** *Ann. Rep. Veg. Res. Stn.,* Wellesbourne, U.K., 1978.

395. **Wang, Da-Yuan and Yan, K.,** Somatic embryogenesis in *Echinochloa crusgalli, Plant Cell Rep.,* 3, 88, 1984.

396. **Wang, P. J. and Hu, N. Y.,** Regeneration of virus free plants through in vitro cultures, in *Advances in Biochemical Engineering Plant Cell Cultures,* Vol. 2, Fiechter, A., Ed., Springer-Verlag, Berlin, 1980, 61.

397. **Wang, D. and Vasil, I. K.,** Somatic embryogenesis and plant regeneration from inflorescence segments of *Pennisetum purpureum* (Napier or elephant grass), *Plant Sci. Lett.,* 25, 147, 1982.

398. **Warnick, D. A.,** *Convolvulus arvense:* rhizogenesis in vitro, *Am. J. Bot.,* 70, 62, 1983.

399. **Weigel, R. C. and Hughes, K. W.,** Long term regeneration by somatic embryogenesis in barley *(Hordeum vulgare)* tissue cultures derived form apical meristem explants, *Plant Cell Tissue Organ Culture,* 5, 151, 1985.

400. **Weis, J. S. and Jaffe, M. J.,** Photoenhancement by blue light of organogenesis in tobacco pith cultures, *Physiol. Plant.,* 22, 171, 1969.

401. **Went, F. W.,** Specific factors other than auxin affecting growth and root formation, *Plant Physiol.,* 13, 55, 1938.

402. **Werner, D. and Gogolin, D.,** Characterization of root initiation and root senescence in callus and organ cultures of *Daucus carota* through determination of the specific activity of the glutamate dehydrogenase (NAD), *Planta,* 91, 155, 1970.

403. **Wernicke, W. and Bretell, R.,** Somatic embryogenesis from *Sorghum bicolor* leaves, *Nature (London),* 287, 138, 1980.

404. **Wernicke, W., Brettell, R., Warizuka, T., and Potrykus, I.,** Adventitious embryoid and root formation from rice leaves, *Z. Pflanzenphysiol.*, 103, 361, 1981.

405. **Wernicke, W. and Milkovits, L.,** Developmental gradients in wheat leaves response of leaf segments in different genotypes cultured in vitro, *J. Plant Physiol.*, 115, 49, 1984.

406. **Wetherell, D. F.,** Enhanced adventive embryogenesis resulting from plasmolysis of cultured wild carrot cells, *Plant Cell Tissue Organ Culture*, 3, 221, 1984.

407. **Wetherell, D. F. and Dougall, D. K.,** Sources of nitrogen supporting growth and embryogenesis in cultured wild carrot tissue, *Physiol. Plant.*, 37, 97, 1976.

408. **White, J. L., Wu, F. S., and Murakishi, H. H.,** The effect of low temperature treatment of tobacco and soybean callus cultures on rates of tobacco and southern bean mosaic virus synthesis, *Phytopathology*, 67, 60, 1977.

409. **White, P. R.,** Potentially unlimited growth of excised tomato roots, *Plant Physiol.*, 9, 585, 1934.

410. **White, P. R.,** Controlled differentiation in a plant tissue culture, *Bull. Torrey Bot. Club*, 66, 507, 1939.

411. **Wicart, G., Mouras, A., and Lutz, A.,** Histological study of organogenesis and embryogenesis in *Cyclamen persicum* tissue cultures. Evidence for a single organogenetic pattern, *Protoplasma*, 119, 159, 1984.

412. **Widholm, J. M. and Rick, S.,** Shoot regeneration from *Glycine canescens* tissue cultures, *Plant Cell Rep.*, 2, 19, 1983.

413. **Williams, E. G. and Maheshwaran, G.,** Somatic embryogenesis factors influencing coordinated behaviour of cells as an embryogenic group, *Ann. Bot.*, 57, 443, 1986.

414. **Winton, L. L.,** Shoot and tree production from aspen tissue culture, *Am. J. Bot.*, 57, 904, 1970.

415. **Wright, M. S., Koehler, S. M., Hinchee, M. A., and Carnes, M. G.,** Plant regeneration by organogenesis in *Glycine max*, *Plant Cell Rep.*, 5, 150, 1986.

416. **Xu, Zhi-hong, Wang, Da-Yuan, Yang, Li-jun, and Wei, Zhiming,** Somatic embryogenesis and plant regeneration in cultured immature inflorescences of *Setaria italica*, *Plant Cell Rep.*, 3, 45, 1984.

417. **Yamaguchi, T. and Nakajima T.,** Effect of abscisic acid on adventitious bud formation in cultured tissues derived both from root tubers of sweet potato and from stem tubers, in *3rd Int. Congr. Plant Tissue and Cell Culture*, Street, H. E., Ed., University of Leicester, England, 1974, Abstr. no. 65.

418. **Yasuda, T., Fujii, Y., and Yamaguchi, T.,** Embryogenic callus induction from *Coffea arabica* leaf explants by benzyladenine, *Plant Cell Physiol.*, 26, 595, 1985.

419. **Zapata, F. J. and Sink, K. C.,** Somatic embryogenesis from *Lycopersicon peruvianum* leaf mesophyll protoplasts, *Theor. Appl. Genet.*, 59, 265, 1981.

420. **Zelcher, A., Soferman, O., and Izhar, S.,** Shoot regeneration in root cultures of Solanceae, *Plant Cell Rep.*, 2, 252, 1983.

421. **Zelcher, A., Soferman, O., and Izhar, S.,** An in vitro screening for tomato genotypes exhibiting efficient shoot regeneration, *J. Plant Physiol.*, 115, 211, 1984.

422. **Zimmermann, R. H.,** Tissue culture of fruit trees and other fruit plants, *Proc. Int. Plant Prop. Soc.*, 28, 539, 1978.

423. **Zimmermann, R. H. and Broome, O. C.,** Phloroglucinol and in vitro rooting of apple cultivar cuttings, *J. Am. Soc. Hortic. Sci.*, 106, 648, 1981.

424. **Zimney, J. and Lorz, H.,** Plant regeneration and initiation of cell suspensions from root-tip derived callus of *Oryza sativa* (rice), *Plant Cell Rep.*, 5, 89, 1986.

Chapter 4

INDUCTION OF HAPLOID PLANT/CELL

I. SIGNIFICANCE

The science of genetics owes its origin to a work on a higher plant when Gregor Mendel, at the turn of last century, based on his simple experiments on the breeding of the garden pea, laid down the laws of inheritance. This youngest science in its applications is, at present, a dominant discipline. Rapid advances in this field have, however, been possible due to work on microbes and the progress can be attributed to their haploid nature, which permits the detection of recessive mutants not possible with a diploid cell or an organism. The importance of haploidy in mutation research can be visualized from the fact that since the discovery of penicillin from a haploid fungus, *Penicillium*, the yield of this drug has increased to more than a 1000-fold, whereas much needed productivity of higher plants which constitute crops for food, feed, fiber, and furnishings has lagged behind. One of the limitations of higher plants is their diploid nature. Hence, haploids of higher plants are highly desirable.

Haploid angiosperms are reckoned to be of great value in plant improvement. A haploid plant on diploidization readily results in a homozygous diploid pure line, otherwise obtained after years of inbreeding. In addition to their applied value, the haploids are of great importance in the fundamental genetics of higher plants. In particular, a haploid cell, since it is not encumbered by problems of dominance and segregation, can simplify the production, identification, and selection of mutants. From a mutant cell line, in turn, plants of desired agronomic characters are possible. Therefore, the work on haploid cells of higher plants is a fine example of usefulness of basic research for applied work. Details about the importance of haploid plants are given at the end of this chapter.

A haploid plant of *Datura* occurring in nature was discovered in the early 1920s.[11] Since then, haploids of natural occurrence have been reported in a number of other plants. Also, it has been possible to induce the formation of haploids. The type of work done has been reviewed from time to time. There are reviews about: (1) haploids of spontaneous origin and induction of haploids in vivo,[24,118,125,136] (2) induction of haploids in vivo and their recovery in vitro,[99,107] and (3) induction of haploids in vitro.[74,137,263]

In this account is given a comprehensive resume of induction of haploids, employing different methods. An extended treatment is given to induction of haploids from pollen grains, on culture of anther or isolated pollen, a method found to be more successful than the rest.

II. NATURAL OCCURRENCE AND INDUCTION OF HAPLOID IN VIVO

A haploid plant of spontaneous origin is parthenogenetic; the embryo develops from an unfertilized egg. Spontaneously produced haploid plants are at times associated with polyembryony. In crop plants this is seen in oil crops: rape, *Brassica napus*[245] and flax, *Linum usitatissimum*.[177,186,246]

Induction of haploid plants in vivo is possible through pseudogamy, development of an unfertilized female gamete after stimulation by male gamete. An important example is the haploid of maize from intervarietal crosses.[22] From interspecific pollination haploids have been possible in potato,[87] lucerne,[10] tobacco,[18] poplar,[222] and wheat.[254]

Less frequently, due to semigamy both male and female gametic nuclei are stimulated to divide forming a mosaic haploid individual, as in cotton.[258] At flowering, some of these plants that were chimeral as seedlings are haploid.

Rarely, haploid plants are possible through androgenesis.[23] The maternal nucleus is eliminated or inactivated before fertilization of egg cell and the haploid individual contains the chromosomes of male gamete alone in egg cytoplasm. The frequency of occurrence of this type of haploid plants is very low, but a dramatic increase in frequency of androgenic haploids was shown in recessive maize mutant "indeterminant gametophyte".[117]

Despite these many methods for the induction of haploids in vivo, the frequency of induction is rather very low and the methods are also time consuming and unpredictable. However, of late, in some plants due to finding of a superior pollinator it has been possible to increase the frequency of formation of haploids. For instance, in 4n × 2n cross of potato employing diploid *Solanum phureja* as a male parent, the frequency of formation of haploids is greatly increased.[148] A superior pollinator is also known in *Zea mays*.[117]

III. INDUCTION OF HAPLOID IN VIVO AND RECOVERY IN VITRO

A method for the induction of haploid in vivo is distant hybridization or alien fertilization. Following fertilization the chromosomes of one of the parents are selectively lost, and the resulting seed from such an embryo is haploid. Employing this method, haploid plants were possible in barley by crossing *Hordeum bulbosum* (2×) as female parent with *H. vulgare* (2×) as male parent.[243] In this way, 59 haploid plants were obtained. However, the reproducibility of results was difficult and many of the haploid plants had characters of *H. bulbosum*, since it was employed as the female parent.

An improvement in technique in terms of increased frequency of haploid formation was possible[108] when *H. vulgare* was employed as female parent and *H. bulbosum* as male parent. Further, in order to prevent the abortion of young embryo the technique of embryo culture was included to recover the seed. By a combination of these two techniques, distant hybridization and embryo culture, induction frequencies of 11[108] and 60%[99] have been possible in *H. vulgare*. The origin of haploid embryo is ascribed to the selective elimination of *H. bulbosum* chromosomes during early stages of development.[108,224] After fertilization, a spray of florets with gibberellin at 75 ppm enhanced the frequency of seed set.[225]

Also, in wheat,[4] haploids have been possible by employing *T. aestivum* as female parent and fertilizing it with pollen of *H. bulbosum*. Here again, chromosomes of *H. bulbosum* are selectively eliminated.

The reasons for selective elimination of chromosomes of an individual are not known. Various explanations advanced by different workers are listed below:

1. It is very likely that the chromosome cycle of *H. bulbosum* is not in synchrony with *H. vulgare*.[203]
2. Chromosome elimination in *Hordeum* hybrids is possibly due to disturbed control of protein metabolism[6] in hybrid seeds. The chromosomes of *H. bulbosum* are selectively eliminated because they are probably less efficient than *H. vulgare* chromosomes in forming normal attachment at spindle proteins.[6]
3. Elimination of *H. bulbosum* in interspecific crosses is a genotypic response.[214] Different genotypes of *H. bulbosum* differ in their ability to produce haploid seeds in barley.
4. It is possibly a temperature-controlled response.[175] At 14.5°C about 50% the embryos were hybrids up to 10 days and this was not affected by increase in temperature to 23°C. By contrast, at 23°C elimination commenced after 1 to 2 days of pollination and this process could not be arrested on transfer to 14.5°C. However, this influence of contrasting temperature, on chromosome elimination, could not be seen in crosses of *H. lechleri* × *H. bulbosum* and *T. aestivum* × *H. bulbosum* whose progeny were exclusively hybrid and haploid,[176] respectively.

The application of these two techniques, distant hybridization in vivo and recovery of embryos in vitro, for the induction of haploids in appreciable frequencies is limited to only two crops, barley and wheat.

IV. INDUCTION OF HAPLOID IN VITRO

Due to the importance of haploid angiospermous plants in fundamental as well as applied research, induction of haploids in vitro, employing the technique of tissue and cell culture, has attracted considerable attention.

To begin with, in view of the parthenogenetic origin of haploid plants in nature, attempts were made to stimulate the development of embryos without fertilization, on culture of unfertilized ovules of angiosperms, but without success. Instead, in a gymnospermous plant it was possible to have differentiation of embryos/plants on culture of female gametophytes of *Zamia floridana*.[126] This was confirmed in *Z. integrifolia*[165] and later also in *Ephedra foliata*.[122] Also, in angiosperms haploid plants have been possible on culture of unfertilized ovaries of barley,[208] wheat and tobacco,[287] and rice.[286] Recently, maternal haploids have also been possible in *Petunia axillaris* on culture of placenta-attached, unfertilized ovules.[41] These results are, however, marked by a low frequency of response.

The origin of a haploid tissue was also sought by the repeated divisions of a pollen. Again, this was also possible initially in gymnosperms on culture of pollen of *Ginkgo*,[255] *Taxus*,[256] *Torreya*,[257] and *Ephedra*.[121] These tissues unfortunately failed to form plantlets. This was soon followed by the reports of origin of haploid plants from microspores of angiosperms.

As for angiosperms, it is a rare incidence that the occurrence of a haploid plant in nature was discovered in *Datura*[11] and the induction of haploid plants in vitro was also possible for the first time in this plant from microspores when anthers of *D. innoxia* were cultured on a nutrient medium.[68,69] This basic discovery of induction of haploid plants from microspores of *Datura* was confirmed on culture of anthers of *Nicotiana*.[163] Since then, induction of haploids has been possible in a number of other plants belonging to different families.[137] More recently, it has also been possible to induce the formation of haploid plants on culture of isolated pollen.[189,192,193] In this account, culture of anthers and culture of microspores for the induction of haploid plants are taken individually.

V. ANTHER CULTURE FOR INDUCTION OF HAPLOID

The anthers of *Datura innoxia*[68] were cultured to study the control of meiosis and instead the emergence of embryo-like structures from these anthers was too much of a spectacle. Later, these embryos were confirmed to be haploid.[69] Since then, this simple technique has turned out to be a very successful method for the induction of haploids in an ever increasing number of angiosperms. Up to 1980, induction of haploids on culture of anthers was possible in 153 species belonging to 52 genera and 23 families of dicots as well as monocots,[138] and within a year, in 1981, this number increased to 171 species belonging to some 60 genera and 26 families of angiosperms.[137]

A. Technique and Response

The culture of anthers for the induction of haploid plants is a simple technique. The anthers can be cultured on agar-sucrose nutrient medium with or without hormones and other supplements. After a couple of weeks, or at the most, a few weeks, some of the microspores in the anther respond to regenerate by forming either embryos/plantlets or divide to form callus tissues. For the sake of clarification, at this introductory stage, it is worth generalizing that microspores of dicots are more responsive and readily form embryos while microspores of monocots are relatively less responsive and can be made to form a tissue which is capable of regenerating into haploid plants.

Although the technique is simple, the response is highly variable. Rarely, it is possible to find a similar response even between anthers from two identical floral buds. Occasionally, all anthers from a bud respond to form haploid plants, whereas anthers from another bud fail to do so. The more common response is one or more anthers from a bud responding to form haploids. Based on their response to form haploid plants/tissues, the plants are classified as responsive-type, low responsive-type, and recalcitrant. Members of Solanaceae readily respond to form haploids and are classified as highly responsive type. However, even in anthers from solanaceous plants the average number of responding anthers ranges from 10 to 30% and rarely exceeds 50%. Also, of the thousands of pollen grains in an anther, very few respond to form haploid plants/tissues. Therefore, for a proper representation the response should be indicated in terms of anther response as well as anther productivity. The former will represent percent responding anthers and the latter, number of haploid plants/tissues per responding anther. However, in most of the investigations, there is information about anther response alone.

B. Factors Affecting Formation of Haploid

Numerous factors have been found to affect the formation of haploid plants/tissues on culture of anthers. These factors range from growth conditions of plants from which anthers are obtained for culture, to physical and chemical conditions prevailing during culture of anthers.

1. Donor Plant

Based on a response from a particular genotype and either its absence or a low response in others, plant genotype is considered to be an important factor affecting anther response and productivity. In highly responsive genus *Nicotiana* (Figure 1A, B) the polyploid species *N. tabacum* is more responsive than diploid *N. sylvestris*.[162,195] Most of *Nicotiana* spp. readily respond on a simple medium, but in *N. langsdorffii*[53] only a few pollen plants were possible even on a special medium during certain growth conditions. Different cultivars of tobacco respond differently;[77] Kupchunos was most productive followed by White burley, Techne, and Badischer Burley. The response in *N. tabacum*[223] was higher in F_1 lines derived from mutants.

Other plants where genotypic differences recorded are *Arabidopsis thaliana*,[65] *Lycopersicon esculentum*,[66] *Oryza sativa*,[70,141] *Vitis vinifera*,[67] *Solanum tuberosum*,[97] *S. chacoense* and interspecific diploid hybrid *S. tuberosum* × *S. chacoense*,[19] *Secale cereale*,[271] *Zea mays*,[248] and *Triticum aestivum*.[127] However, it is not clear whether genotypic differences are the representation of genetic variability or are a reflection of improper culture conditions.

The genetic control of response can be seen in breeding experiments aimed to accumulate genes favoring haploid formation. In *S. cereale*[271,273] and *S. tuberosum*,[98] intercrossing resulted in cultivars more responsive to haploid production than the parental types. An additional support can be seen in a work on tomato[283] where a single gene mutation could increase the frequency of pollen callus formation and plant regeneration. Also, in *Petunia*[146] anther response could be manipulated by varying the genotype. From backcrosses between *P. hybrida* Rose du Ciel and *P. axillaris*, individual plants were obtained whose anthers formed plantlets at rates greater than 5%. Among monocots, in wheat[174] a higher frequency of pollen plants was possible in microspore-derived lines than in the controls. There is evidence that technique selects against certain types of microspores. An analysis of pollen plant in pearl millet[16] obtained from hybrid, Massue ♀ × ligui ♂, indicated that in agronomic characters they were close to ligui parent. Also, heritability of callus formation and plantlet regeneration has been shown in wheat.[127] A high frequency of callus formation was seen in anthers of kitt × olaf cultivars, whereas a higher mean plantlet formation was possible in anthers of cultivar "Fielder". In *Oryza sativa*[141] callus formation from microspores on culture

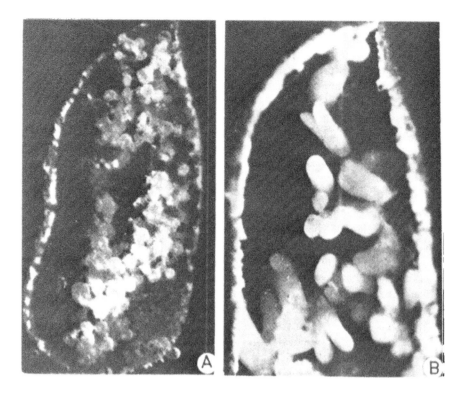

FIGURE 1. Embryo formation on culture of anthers of *Nicotiana tabacum*, a highly responsive genus. (A) An anther split open to show globular embryos. (B) Later stage showing well-developed embryos. (From Nitsch, J. P., *Phytomorphology*, 19, 389, 1969. With permission.)

of anthers has been shown to be a recessive character conditioned by a single block of genes. Transfer of such genes to elite germplasm via crossing may be the most effective way for realizing the potential advantage of haploids.

Although there is a pronounced effect of genotype on anther response, low response or no response of a recalcitrant type is to be seen in the context of physiological conditions employed for culture of anthers. Unresponsive tobacco species[162] responded to form haploid plants,[249] and the lower response of a particular cultivar could be increased by a change in light condition during anther culture.[171] Similarly, nonresponsive[65] *Arabidopsis* spp. could be made responsive[1] by a change in physiological condition of donor plants. Therefore, it is very likely that genotypes differ in their physiological requirement to form haploids.

The age of the donor plant is another important factor affecting anther response. Generally, the buds appearing in the beginning of the flowering season are more responsive than later formed buds. This has been shown in *Datura metel*,[153] *Nicotiana tabacum*,[229] and *Atropa belladonna*.[194] The reduced response is ascribed to the decline in pollen viability[229] and the decline in vigor of donor plants particularly during seed set.[158] In *Datura*,[158] the effect of plant age could be overcome by prevention of seed set on removing the older buds. However, in some plants such as *N. tabacum*[102] cv. Havana there is no effect of plant age.

Seasonal variation in anther response, a reflection of change in the physiological condition of donor plant, has been reported in *Solanum tuberosum*,[48] *Hordeum vulgare*,[38] and *Triticum aestivum*.[173] The change in physiological condition of a donor plant and its consequent effect on anther response can be better understood when results of plants kept in controlled conditions are considered. The effect of changing photoperiod can be seen in *N. tabacum* cv. White Burley.[44] When parent plants were grown in an 8-hr photoperiod at 14,000 lux, 56%

of the anthers responded to form 19 haploids per anther as compared to 35% response from anthers taken from plants grown in a 16-hr photoperiod of the same light intensity.

A lower growth temperature of donor plants is favorable for haploid formation in *Brassica napus*;[116] the donor plants grown at low temperature gave greater embryo yield. This has been confirmed by other investigators.[47,247] In an intensive study[247] it was found that plants grown at 20 to 15°C unshaded in a daylight phytotron were most productive. Further exposure to 4 weeks of vernalization was accompanied by a significant increase in response. Also, in *N. tabacum* var. Badischer Burley[190,191] there was an increased response from anthers taken from plants flowering at 15°C than at 25°C. By contrast, in *N. knightiana*[178] the anther response doubled by an increase in plant growth temperature from 14 to 20°C. Also, in barley[59] and wheat[264] an increased response was possible from anthers derived from plants growing at an increased temperature during summer.

Certain treatments to donor plants also affect anther response. An application of 2-chloroethylphosphonic acid (ethephon, an ethylene releasing compound) to wheat plants resulted in division of pollen grains while still in the spike.[5] Also, in *Oryza sativa*[265] ethephon application for 48 hr at 10°C to inflorescences resulted in increased anther response. However, applications of etherel had no effect in *Capsicum annuum*.[135] Also, etherel, cycocel, or GA_3 had no effect in *N. tabacum*.[76] Instead, application of NAA shifted the sex balance of flowers towards femaleness and increased pollen plant formation on anther culture.

Other treatments to either plant part or whole excised plant, causing stress, also affect anther response. In wheat,[172] removal of the apical region of spike increased the frequency of microspores having identical nuclei which is considered to be conducive to haploid formation. Also, in barley[278] frequency of pollen with identical nuclei and incidence of callusing could be increased when plants were clipped at the ground level and maintained in water for 1 to 2 days before the culture of whole spike. A condition of stress induced by nitrogen starvation[231] to donor plants of *N. tabacum* cv. White Burley resulted in an increased anther response as well as anther productivity. However, this was not true for *N. tabacum* cv. Badischer Burley where optimal nutritional condition in terms of mineral supply to the growing plants gave a better response.[78] These two contrasting results in the same plant can be explained if we consider other growth conditions. White Burley plants were grown in a greenhouse, whereas Badischer Burley plants were grown in precise controlled conditions. Nitrogen starvation had no effect when plants were grown at 24°C. However, when plants were grown at 15°C nitrogen starvation resulted in an increase in pollen plant formation.[76]

Therefore, keeping in view the variable response in anther culture and growth conditions of donor plants affecting the response, for reproducible results, the donor plants should be grown under controlled conditions. Anthers from flowers formed in vitro on thin-layer explants of *N. tabacum* and *N. plumbaginifolia*[251] were more productive than anthers from plants flowering in vivo.

2. Anther/Pollen Stage

Since a haploid plant/tissue originates from a pollen, a critical factor for success is the proper stage of pollen at the time of anther culture. To determine the required stage of pollen, staging of buds and anthers is done by examining one anther per bud while the rest are retained for culture. In view of prevailing synchrony in development of microspores in an anther,[56] this is the standard practice, but significant differences[262] are known to occur in the stage of development of pollen within a single anther or between different anthers of a flower bud. Hence, frequent checking is desirable, particularly when plants are not grown under controlled conditions. The staging of pollen is quite simple, but it is error prone and at times can be difficult and even misleading. For emphasis, a brief account of the developmental sequence of a microspore to a pollen is in order.

At the microspore stage there is sparse cytoplasm and it is revealed by a weak staining reaction towards acetocarmine. This is followed by a stage of vacuolation and as the vacuole develops the nucleus from its central position is displaced towards the periphery. At this stage the nucleus undergoes DNA replication, increases in size, and there is simultaneous synthesis of RNA and proteins.[140] Following this, an asymmetric division of microspore results in a smaller generative cell with a densely staining nucleus and a large vegetative cell with a diffuse nucleus. To begin with, the generative cell is attached to the intine, but ultimately it becomes free and the unique situation of a cell within a cell is established. In some genera, particularly those belonging to Cruciferae and Gramineae, the generative cell divides further to produce two male gametes, whereas in most other dicots this division is delayed up to the germination of pollen.[15]

The misleading factor in the staging of pollen is the variable rate of cytoplasmic synthesis in vegetative cell. This variability can be demonstrated in two solanaceous genera, *Nicotiana* and *Datura*. In *N. tabacum*, before the detachment of generative cell from intine, there is a rapid synthesis and the vegetative cell is filled with cytoplasm; it shows dense staining with acetocarmine, its diffuse nucleus is often obscured, and the pollen can be mistaken to be uninucleate. By contrast, in *Datura innoxia*, due to slow cytoplasmic synthesis and the consequent poor staining with acetocarmine, the young pollen grain, with vegetative and generative nuclei, may be mistaken as microspore.

The formation of pollen plants has been possible on culture of anthers at different stages of pollen development, ranging from tetrads to tricellular pollen. However, in most of the investigations unicellular stage of pollen is found to result in an optimal response, an increased number of haploids in dicot[194,195] as well as monocot[73,167] plants.

The stage of pollen is not only a critical factor for obtaining pollen plants, but also determines their ploidy. In *Datura*[55,153] and *Nicotiana*[54] plants from uninucleate pollen were mostly haploid, whereas plants from anthers at a later stage of pollen had a higher chromosome number. Also, in *Hyoscyamus niger*[36] haploid plants were possible only from uninucleate pollen. In *D. innoxia*[235] it has been confirmed that 80% of the plants from binucleate pollen were nonhaploid. Further, in *Datura*[35] it has been found that not all pollen are haploid. In the first-formed flower buds about 1% of the pollen are unreduced, and this frequency increases with the age of plant so by the end of the flowering period it may be as high as 10%. Unreduced microspores are also described in *Secale*.[271] Plants of *Datura innoxia*[234] when grown at temperature lower than 20°C show meiotic irregularities leading to formation of nonhaploid microspores.

The staging of anthers based on developmental stage of pollen becomes questionable when we consider the fact that only a small number, rarely exceeding 100, of pollen among the thousands present in an anther produce pollen plants. An answer to the question why only a fraction of pollen is responsive can be seen in studies on anther cultures of barley[37] and tobacco.[86] In these plants pollen capable of forming embryos can be distinguished from the majority, destined to form gametes. The embryo-forming pollen are small, with a cytoplasm that stains less intensely with acetocarmine than the gametophytic pollen. Also, in wheat,[244,285] embryos arise from smaller pollen on anther culture. The presence of smaller embryogenic pollen and bigger gametophytic pollen is defined as pollen dimorphism and the significance of this phenomenon in relation to pollen plant formation has been recently reviewed.[188]

3. Pretreatment of Anther/Bud

A cold pretreatment of buds of *Datura innoxia*[161] by keeping them in a refrigerator for 48 hr resulted in a dramatic increase in the number of pollen plants on culture of anthers. Since then, cold treatment has been found to be effective in almost all plants tried so far and is considered to be an important factor controlling pollen plant formation. Cold treatment is also effective when it is given to anthers on culture.[43,159,242] However, it is not as effective as given to anthers while they are in flower buds.[159,242]

Cold treatment has been given at different temperatures and for variable periods. Therefore, different authors, possibly depending upon the temperature employed, have described the treatment as chilling, cold treatment, and even stress treatment. For floral buds of *D. innoxia*[156] a chilling treatment at 3°C for 48 hr has been recommended; however, in a detailed study, 4°C for 4 days was found to be optimal.[259] In *N. tabacum*[157] a treatment at 5°C for 72 hr has been recommended, but others have found that optimal treatment is 7 to 8°C for 12 days[239] or 10°C for 10 days.[189] The optimal dose in *Hyoscyamus niger*[242] is 15°C for 5 days and it is 4°C for 21 to 28 days in *Hordeum*.[90] In a recent study, highly significant interaction is shown between different periods of cold treatment at different temperatures between different genotypes of *N. rustica*[28] in terms of anther response, anther productivity, and days to first plantlet formation. Therefore, for every plant an optimal dose of cold treatment has to be worked out.

The method for administration of cold treatment is also variable from laboratory to laboratory. The large buds of *Datura*[259] can be held erect with their cut ends dipping in drops of water and were kept in cold in this way after enclosing the container in aluminium foil. The same could be done with buds of *Nicotiana*.[189] An alternative[236] method employed for tobacco buds, claimed to be easier, is to enclose them in a polythene bag containing a few drops of water. The bags after sealing are kept in the refrigerator in dark. In yet another variable way, whole tillers of barley[280] were put in a petri dish, awns removed, the dish was sealed with parafilm, and placed in dark at 7°C for 14 days. Further, it was found that pretreatment of excised spikes were more effective than pretreatment of tillers.[90] Pretreatment of spikes at 4°C was more effective than at 7, 14, or 20°C; 3 to 5 weeks were required for maximal yield. In sugarcane;[57] a long cold treatment of panicle branches, 10°C for 4 to 10 weeks in low salt media was effective. Variation in intensity and duration of cold treatment significantly affected the response in wheat.[128] For an optimal response, 2 weeks of treatment of cut tillers at 4°C in dark was required.

The precise function of cold treatment and its mode of action remains to be worked out. It is essential to do so because cold treatment is a key determinant in the induction of pollen plants. An understanding of this process is likely to be rewarding. In the initial work on *D. innoxia*[161] it was proposed that cold treatment stops the existing metabolism and if the period is long enough, it will induce the pollen towards a new metabolism which is conducive for division of pollen on culture of anthers. Further, it was proposed[156] that cold treatment increases the frequency of pollen plant formation by altering the polarity of pollen in such a way that when given the stimulus of culture they divide by an equal, instead of an unequal division. Also, cold treatment was ascribed to support an increased frequency of viable pollen on culture of anthers. Besides cold treatment, in *Datura innoxia* centrifugation[205] of buds prior to culture of anther also resulted in a slight increase in response, but the effect was accentuated when centrifugation was combined with cold treatment. Centrifugation of pollen promotes callus formation,[13] it results in disorganization of microtubules and cold treatment leads to their dissolution.[81] Therefore, it is very likely that in *Datura* an interference in polarity of pollen leads to increased response. However, in tobacco[43] cold treatment supported only an increased frequency of viable pollen and did not result in an increased frequency of pollen with identical nuclei.[43,232,240] Also, in barley,[237] where cold treatment is effective, it does not result in increased frequency of pollen with identical nuclei. However, in a recent study, an incubation of panicle sections of *Saccharum spontaneum* at 10°C for 3 weeks resulted in a loss of viability of a large number of pollen. Of those that survived a great many divided symmetrically rather than asymmetrically.[83]

A new role that has been ascribed to cold treatment is of stress treatment. In barley[233] it is proposed that cold treatment removes the constraints to division of pollen by way of desynchronization of pollen and tapetum and this results in the repression of genes essential for gametophyte development. An evidence in favor of desynchronization was seen in freeze-fracture study of cold-treated anthers. During cold treatment there was premature dispersal of tapetum.[238]

Another new role of cold treatment that has come to light is the reduction in the concentration of abscisic acid (ABA). In *Anemone canadensis*,[104] where cold treatment (an optimum of about 20 days) was stimulatory, it reduced the concentration of ABA from 2.2×10^{-6} g/g anthers to 0.6×10^{-6} g/g anthers.

In an ultrastructural study of *Datura* pollen,[207] after cold treatment (3°C for 48 hr) of flower buds, some reversible changes associated with the treatment were (1) vacuolation, (2) loss of cytoplasmic structure, (3) disorganization of organelles, (4) darkening of cell wall, and (5) apparent loss of turgor pressure. However, a causal relationship could not be established between these changes and haploid formation.

The above explanations about the role of cold treatment are not consistent and hence need clarification and confirmation. However, an unequivocal role of cold treatment is the delay in senscence[169] of anthers. This positive role of cold treatment can be understood from the fact that rapidly senescing anthers do not form haploids[143,169] and they also reduce the frequency of haploid formation in responding anthers.

Interestingly, in anther culture of *Annona squamosa* chilling of flower buds had no effect; instead, a new pretreatment of anthers was found to be essential for the formation of haploids. This essential treatment[150] included dissection of flower buds in 6% sucrose and 0.25% activated charcoal solution prior to culture of anthers.

4. Posttreatment of Anthers (on Culture)

Isolated anthers, on culture, can also be given a cold treatment for a short period and this also results in an increased response.[43,159,242] However, it is not as effective as cold treatment given to anthers in buds prior to culture.

More interesting is the obligatory requirement of high temperature treatment to anthers of *Brassica* on culture for the formation of haploid plants. If the anthers are cultured continuously at 25°C there is no response. By contrast, a high temperature treatment at 30°C for the first 14 days of culture in *B. napus*[113] and at 35°C for 1 to 3 days in *B. campestris*[114] resulted in induction of haploids. Also, in rice[267] a brief high temperature treatment to anthers, on culture, results in increased frequency of formation of haploid plants. This is also true for wheat[289] and hybrid of *Triticum aestivum* × *T. agropyron*. For these plants temperature treatment is beneficial, but not obligatory.

It is very difficult to comprehend the obligatory requirement of high temperature treatment to *Brassica* anthers, particularly when an optimum response is possible from anthers taken from plants grown at low temperature. In *B. juncea*[63] a combination of cold treatment (10°C for 3 days) and high temperature treatment (31°C for 4 days) was essential for induction of haploids on culture of anthers. By contrast in *B. alba*[130] (= *Sinapis alba*) neither a cold treatment to buds nor a heat treatment to cultured anthers was required. This species seems to be exceptional because for other species of *Brassica*, the need for heat treatment has been repeatedly emphasized. In *B. napus*[47] there is an absolute requirement of 35°C treatment for 2 days for the induction of embryos. Any pretreatment completely inhibited the response. However, plants grown at 15°C gave more productive anthers than at high temperature.

5. Anther Orientation

Of late, anther orientation has been found to be an important factor affecting the response in barley.[92,210] Barley anthers from cold-treated spikes produced zero to a few pollen calli when cultured flat with both loculi in contact with the medium. However, when anthers were cultured in up position, one locule in contact with the medium, there was an increased frequency of responding anthers and they were more productive.

6. Anther Wall

The anther wall can be safely described as having a beneficial role because haploids have been readily possible in a number of plants, albeit at a low frequency, on culture of pollen within the confines of anthers. In highly responsive genera, like *Nicotiana*[162] and *Datura*,[230] the induction of division in pollen is possible if anthers are cultured on a simple agar-sucrose medium, indicating thereby that the initial nutrient requriements are possibly met by anther wall. In *Nicotiana*[169] placing of anthers in a humid environment alone is sufficient to induce division in pollen grains.

In contrast to its beneficial role, the anther wall can also be inhibitory to the initiation and development of pollen plants. This is particularly evident in rapidly senescing anthers.[143,169,259] In *Datura* a serial transfer of anthers to fresh medium at regular intervals resulted in increased frequency of response, indicating that on transfer of anthers from one medium to another there is dilution of inhibitory factor.[259] Senescing anthers not only fail to produce pollen plants, but also retard the frequency of haploid plants from anthers. Therefore, the anther wall is both promotory as well as inhibitory to initiation and development of pollen plants. In those plants which fail to respond, it is very likely that the inhibitor level is high in the anther wall.

Factors other than the anther wall may be responsible for the low yield of haploid plants. This is evidenced in microspore culture of *Saccharum spontaneum*,[84] where calli produced from isolated pollen ceased development at about the same stage as did the calli in an anther.

7. Pollen Viability/Degeneration

A new factor affecting haploid formation described in recent investigations is pollen viability, or pollen degeneration. In slow-browning anthers of *N. tabacum* which result in the production of haploid embryos, with an increase in frequency of degenerating pollen there was a corresponding increase in frequency of embryos.[144] Pollen embryogenesis is possibly promoted by substances released from degenerating pollen. Also, in *Datura* pollen degeneration is recorded, but its relationship with pollen plant formation is to be established. During the first week of anther culture of *D. innoxia*, the frequency of viable pollen decreased progressively, reaching 18% after 5 days of culture. The surviving pollen were uninucleate microspores and pollen with two identical nuclei initiated in vitro. Embryos developed from these two types of pollen.[206]

In wheat[79] there is also a high mortality of pollen on culture of anther during the first week and it is followed by a latent period of about 1 week, and then embryogenic mitosis commences. However, in this plant, instead of total degeneration, the rate of degeneration seems to determine the response. Of the two genotypes, no embryos were produced in genotype A, whereas in genotype B, at least one embryogenic anther was there per half of cultured spike. After 7 to 9 days only 25% of the pollen were viable in genotype A, whereas in genotype B, it was 50%. After 17 days only 2% of pollen were viable in genotype A and it was 18% in genotype B. Therefore, in wheat the genotypic difference observed in terms of anther response does not seem to be linked to induction process, but is linked to different rates of pollen degeneration.

From these results on anther culture of tobacco, datura, and wheat, it is not clear how pollen degeneration affects haploid formation. This is to be looked into in greater detail, particularly when degenerating pollen have been shown to have a deleterious effect. It is seen in experiments on microspore cultures that embryo formation was possible on removal of nonviable microspores.[84,189]

8. Chemical Environment of Culture

Except for the highly responsive genus *Nicotiana*, where initiation of pollen division is possible in a humid environment,[169] in all other genera worked out so far at least mineral elements and sucrose are required. For some genera hormones and other additives are also essential components of nutrient medium. On the basis of results in different genera a distinction is possible. In dicot plants and, particularly, in members of Solanaceae the entire process from initiation of division of pollen to the formation of pollen plant is possible on a simple mineral-sucrose medium, whereas monocot plants in general require hormone-enriched medium for formation of haploids and the response can be further increased by inclusion of undefined complex substances like coconut milk and potato extract, etc.

a. Mineral Elements

The mineral requirement for the culture of anthers is met on employing any one of the different nutrient formulations.[12,134,149,163,277] Of these, most of the formulations were originally designed for the culture of cells and only one[163] has been worked out specifically for the culture of anthers. This is the modification of Murashige and Skoog (MS) medium,[149] in terms of an altered ammonium concentration and deletion of minor elements and vitamins. The minor elements and vitamins do not seem to be essential for pollen plant formation in many members of Solanaceae. It is customary to employ a high salt medium rather than a low salt medium. However, occasionally there is some response on a particular medium and it is altogether missing on another medium. In *Atropa belladonna* there was good response on Linsmaer and Skoog (LS) medium, but it was negligible on Nitsch's medium.[284]

A systematic effort is desirable to identify the influential components of the medium. In *Atropa belladonna*[194] and *Datura*,[218] iron in the form of chelate Fe-EDDHA was better than Fe-EDTA. In monocots, barley[34] and rice,[32] a higher concentration of ammonium inhibited the formation of callus tissue from pollen and, consequently, a new formulation, N_6 medium, has been found to be more effective than earlier formulations employed for anther culture of cereals.[62,164,226] A high nitrate/ammonium ratio also favored haploid formation in the anther culture of sugarcane.[57]

b. Carbohydrate

Customarily, sucrose is the source of energy in a nutrient medium. For most of the dicot plants a low level of 2 to 3% sucrose is quite sufficient for induction of division in pollen and formation of pollen plants. However, in certain dicots, *Brassica campestris*,[115] *B. napus*,[52] and *Solanum tuberosum*,[216] the culture of anthers initially on high sucrose medium is beneficial for induction of division in pollen, but anthers are to be transferred to a low sucrose medium for formation of pollen plants, whereas for monocots in general, a higher level of sucrose (6 to 12%) is required from the beginning to the end of the culture.[29,32,226] The best response in maize is possible at 12 to 15% of sucrose.[123,145] The requirement of a high level of sucrose in anther culture, in comparison to a low level required for cell culture, needs to be explained. There is evidence for and against the suggested osmoregulatory role.[9] In *S. tuberosum*[216] a low level of sucrose along with mannitol was not as effective as a high level of sucrose. By contrast, high osmolarity enhanced pollen callus formation in *Triticale*.[26]

An alternate source of energy in the form of other disaccharides was not as effective as sucrose in *Brassica*[115] and *Datura*.[205] Of the monosaccharides, glucose was effective to an extent,[271] but not fructose.[115] However, in *Petunia*, except for galactose, other sugars tried (sucrose, lactose, maltose, glucose, and fructose) were effective in supporting the formation of haploid plants. It is quite interesting[204] that the formation of haploid was possible on lactose, and maltose gave better results than glucose and sucrose.

c. Hormones

The requirement of hormones for the initiation of division in a pollen to the formation of a pollen plant remains to be understood. The entire process is possible on a hormone-free medium in many plants, particularly those belonging to Solanaceae: *Nicotiana tabacum*[162]

Datura innoxia,[215] *Atropa belladonna*,[196] *Hyoscyamus niger*,[36,179] and *Saintpaulia ionantha*.[91] However, in a difficult-to-respond solanaceous genus, such as *Lycopersicon*,[66] and others such as *Arabidopsis*[65] and *Vitis vinifera*,[67] the formation of a tissue from a pollen followed by its differentiation into haploid plantlets is possible if the anthers are cultured on an auxin-cytokinin medium. Auxin as well as cytokinin are also essential for anther culture of *Citrus aurantifolia*.[25] On this medium, pollen could be stimulated to form embryos. By contrast, auxin alone is essential for success in the anther culture of *Digitalis obscura*.[170]

On a hormone-free medium there is practically no response in most of the cereals worked out so far. For a response, that too of a low order, auxin is required for rice,[155] barley,[34] and sugarcane.[57] On culture of anthers, the pollen grains are stimulated to divide and form a callus which in turn differentiates into haploid plants on an auxin-free medium. On the other hand, for wheat[167] both auxin and cytokinin are required, but differentiation is possible on a hormone-free medium. In this context it is interesting to find that maize[160] is an exception. Haploids in maize are possible without hormone, and antiauxin triiodobenzoic acid (TIBA) enhances embryo formation from pollen grains.

Since hormones are known to bring about cytological abnormalities, their use should be restricted unless it is obligatory. Hormones promote the response even in those genera which readily form plantlets on a basal medium. Cytokinin is stimulatory for *Datura*.[68,69,221] Auxin is stimulatory for *Nicotiana tabacum*,[162] *Lycopersicon esculentum*,[39] and *Hyoscyamus niger*.[181] In *H. niger*, auxin 2,4-D stimulated division even in those pollen grains which would have remained quiescent.

d. Nitrogenous Compounds

As for specific organic nitrogenous compounds promoting haploid formation, glutamine is stimulatory in many dicot plants, whereas it is asparagine for potato[268] and proline for maize.[160]

e. Undefined Substances

Among monocots, wheat is particularly known for a very low response, even on a hormone-rich medium. Nonetheless, a simple 20% aqueous extract of potato supported an increased frequency of response. The variable results on this medium, depending upon the potato cultivar employed to prepare the medium, led to the modification of medium which in addition to potato extract was enriched with major salts, iron, and thiamine. This medium has been found to be quite effective for induction of haploids in *Triticum*.[33] Potato extract medium is also effective for other monocots. In barley its efficacy can be improved on a conditioning of the medium by the preculture of anthers.[280] The conditioning of medium, for culture of barley anthers, cv. Sabarlis, could be done by ovaries and they were more effective than anthers, whereas glumes and other parts of spike were ineffective. Anthers of wheat, rye, oat, maize, tobacco, and other genotypes of barley were not as effective as anthers of Sabarlis. The conditioning factor (CF) could be partially replaced by inclusion of inositol[281] into the medium at 1000 mg/ℓ, whereas glutamine had no effect. Inositol and CF acted synergistically. The effect of CF and inositol was destroyed by heat.

An undefined substance, like potato extract, and a conditioned medium are quite stimulatory for pollen to form pollen plant/tissue, but it is difficult to know about their active ingredients. For want of this information, induction of haploids in cereals is more of an art than a science. A beginning towards the identification of effective factor in conditioned medium has been made.[120] The most effective conditioned medium for barley was preculture of ten ovaries per milliliter of medium for 7 days. This medium supported the division of isolated microspores, regeneration of calli, and formation of plantlets. Gel filtration of conditioned medium revealed two molecular probes having conditioning effect. There is also qualitative and quantitative change in amino acids and sugars of conditioned medium.

In particular, one compound detected by thin-layer chromatography, in the conditioned medium, supported the growth of microspores as compared to unconditioned medium.

9. Physical Environment of Culture

Of the various parameters within the physical environment of culture, the importance of temperature has already been emphasized in an earlier section (V.B.4.) Barring the exceptional *Brassica*, which requires an obligate high temperature treatment during initial period, in most of other genera, such as *Nicotiana*[229] and *Datura*,[220] an optimal response is possible if anthers on culture are maintained at 25 to 30°C. At 30°C in rice,[89] although induction of pollen callus is favored, the callus shows a reduced potential for differentiation into haploid plants. Also, if the cultures are maintained at 30°C there is more of an increase in the number of albino plants[7,266] than at 25°C.

About the role of light and its duration, there are contradictory results. An alternating light-dark sequence was beneficial for anther cultures of *Nicotiana tabacum*,[229] *Hyoscyamus niger*,[36] and *Datura innoxia*.[220] Continuous light was employed for *D. metel*,[153] whereas it was inhibitory for *D. innoxia*[220] and *Anemone virginiana*.[102] Continuous light greatly increased the frequency of pollen embryos in some cultivars of *Nicotiana*.[171] For *N. sylvestris*, 300 lux of continuous light was beneficial for anther response as well as anther productivity.[212]

Other physical factors described affecting pollen plant formation in anther cultures are hydration and gaseous atmosphere. Interestingly, two contrasting situations, water stress as well as water saturation during initial period of anther culture, are described to be promotory in *N. tabacum*. Short-term water stress condition imposed on anthers by adding 0.5 M mannitol to the medium stimulated the response. Water stress for 1 day resulted in enhancement of endogenous ABA level from 0.059 ng/grain to 0.120 ng/grain. When ABA was added from outside, 3-day treatment at 10^{-5} M, it also stimulated the response.[93,95] By contrast, water saturated atmosphere for 2 to 3 days enhanced the anther response to fourfold and doubled the anther productivity.[46] It is difficult to comprehend these two contrasting results in the same genus. However, it remains to be seen whether this difference is typical of different cultivars employed for these studies. A change of gaseous atmosphere is also stimulatory. Short-term (30 to 60 min) anaerobic atmosphere by treatment with 100% N_2 stream at 1 bar promoted haploid formation in tobacco anthers.[95] The treatment, with N_2 containing 2.5 and 5% O_2 also had stimulatory effect. Also, elevation of CO_2 concentration to 2% promoted haploid formation in different species of *Anemone* and *Papaver*.[103,104]

The physical environment of the culture can be modified in an indirect way by the inclusion of substances which are not metabolized by the tissue, but the presence itself affects the response. Charcoal is one such substance. Although charcoal can not be taken up by anthers, its inclusion in agar medium dramatically increases the response in many plants tried so far such as tobacco,[2,151] rapeseed,[271] potato,[217] maize[145] *Triticale*,[226] and *Datura*.[260] At present, petunia[139] is the only exceptional plant where charcoal is ineffective in promoting haploid formation. Barring this exception, the promotory effect of charcoal is possibly due to adsorption of inhibitors present in agar as well as released from senescing anthers.[274] It can also adsorb inhibitors[269] arising in medium during its preparation, such as 5-hydroxymethyl furfural, a product of sucrose degradation during autoclaving. Therefore, in *Datura innoxia* charcoal containing agar medium was better than liquid medium (Figure 2) prepared by autoclaving.[260]

Inclusion of charcoal in liquid medium is inhibitory to response. Instead, floating of anthers of *Anemone, Clematis, Papaver*, and *Nicotiana* on liquid medium which overlay an agarified charcoal medium was superior to conventional methods — culture either on solid or liquid medium.[104] This method of anther culture was effective in anther cultures of *N. paniculata* and *N. sylvestris*.[212] Employing double layer medium, upper layer of liquid medium and lower layer of agar medium containing activated charcoal, it has been possible

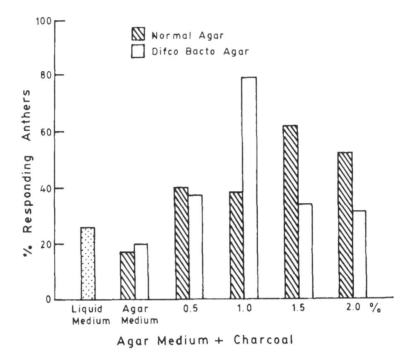

FIGURE 2.   Enhancement of pollen embryo formation by charcoal on anther culture of *Datura innoxia*. (From Tyagi, A. K., Rashid, A., and Maheshwari, S. C., *Physiol. Plant.*, 49, 296, 1980. With permission.)

to resolve[101] that (1) removal of activated charcoal before the inoculation of anthers also stimulated embryogenesis. This clearly indicates that charcoal removed inhibitors present in the medium. (2) Medium without charcoal when inoculated with anthers had a high concentration of phenolics which could be reduced to one fifth on inclusion of charcoal. (3) The stimulating effect of charcoal was most pronounced when first visible embryos had emerged. However, a caution is necessary in using charcoal, particularly when the medium employed is enriched with hormones or chelates. Charcoal adsorbs these substances.[75,139,269]

Inclusion of a polymer polyvinylpolypyrrolidone, into the medium has also been found to be beneficial for anther culture of *Datura*.[3,261] The effect is possibly due to the adsorption of phenolics emanating from anthers, which inhibit the response.

Other factors, such as the number of anthers per culture vessel,[60,142] positioning of anthers on agar medium,[220] and culture vessel atmosphere,[45] affect the response.

C. Anther Culture — An Overview

The culture of anthers on agar medium is a simple and successful technique, but it has certain limitations. The main limitation is a low and variable response. One of the reasons for reduced response can be inhibitory factors present in anther wall. In anther culture it is also difficult to rule out the possibility of a mixed population of haploid and diploid plants, particularly when in some plants hormones are essential for induction and formation of pollen plants.

A number of factors affect the response in anther culture. Therefore, it is not only difficult to understand the mechanism of action of these factors, but it is also difficult to define their mode of action — whether they act individually or in an interacting manner. A significant interaction between genotype and growth conditions of donor plants is seen in *Nicotiana*[76,77] and *Brassica*.[247] Also an interaction is seen between genotype and low temperature pretreatment to anthers of *N. rustica*.[28]

D. Refinements of Anther Culture

In view of the limitations of anther culture on agar medium, attempts made to modify the technique have turned out to be rewarding.

A mere change in the culture of anthers from agar to liquid medium resulted in increased frequency of pollen embryos in *N. tabacum*[274] and *Brassica napus*.[132] Further, in a modification of anther culture on liquid medium it was possible to have pollen plant formation outside the anthers in *Nicotiana* and *Hyoscyamus*.[275] For this, after an initial period of culture of anthers of *Nicotiana* for 4 days and *Hyoscyamus* for 5 days, a slit along the line of anther dehiscence followed by gentle squeezing of anthers resulted in suspension of pollen, outside the anther, which formed pollen plants at a much higher frequency.

Simultaneously, an attempt on anther culture of *Nicotiana* in liquid medium resulted in formation of pollen suspension, after 6 days of culture, due to spontaneous dehiscence of anthers.[239] In this work, 12-day-cold-treated anthers were employed. Further, a transfer of anthers to fresh medium resulted in another fraction of shed pollen. Following this, a simple expedient of serial subculture of anthers at regular intervals resulted in several fractions of shed pollen. The formation of embryos in all fractions of shed pollen, without the presence of anther, led the authors to describe this attempt ''a new approach to pollen culture,''[239] and later it was remarked that ''such cultures for all intents and purposes are pollen culture.''[232]

The success of serial culture of anthers in liquid medium, resulting into an increased response, lies in the dehiscence of anthers. It can either be spontaneous or due to growth conditions of donor plants or treatments of anthers prior to culture. Anthers of *Datura innoxia*[232] from greenhouse plants do not dehisce in liquid medium, but anthers from garden grown plants do so, and transfer of anthers to new medium at regular intervals results in serial cultures of shed pollen.[259] In anthers with a prior cold treatment there is not only an increased shedding of pollen, but also an increased number of plantlets in different fractions. A higher frequency of embryo formation in different fractions of serial culture (Figure 3) can be understood from the fact that it minimizes the disadvantages of anther culture on agar medium and maximizes its advantages. Also, in serial cultures of anthers, the pollen and developing pollen plants are shed in a medium which is preconditioned for their growth by the presence of anthers and has the minimum of inhibitors elaborated by anther wall because anthers are frequently transferred to new medium. In addition, the developing embryos are not only released in a more favorable nutrient milieu, but they are also liberated from the limitation of anther space. Because of these advantages and ease of manipulation, serial culture of anthers are more productive. In instances where anthers do not dehisce spontaneously, even after cold treatment, surgical splitting of anthers can be tried and for dispersal of pollen, light agitation for some period can be introduced.

A modification of liquid culture medium is possible by inclusion of ficoll. A high frequency response was obtained from barley anthers on ficoll medium.[106] Ficoll helped the pollen calli float on the surface ensuing better aeration and growth of calli.

E. Origin of Pollen Plant

Details about the origin of the pollen plant have been worked out in a number of plants by different workers.[180,194,195,215,237,241] To begin with, a multicellular structure arises from a microspore or a pollen, and this is possible in one of the following four ways:

1. A microspore divides directly, leading to the formation of two identical cells; both these cells participate in the formation of a multicellular structure.
2. In a pollen, with a vegetative cell and a generative cell, divisions of vegetative cell alone result in the multicellular structure.
3. Division products of generative cell as well as vegetative cell participate in the formation of the multicellular structure.
4. Generative cell and its derivatives alone result in the multicellular structure.

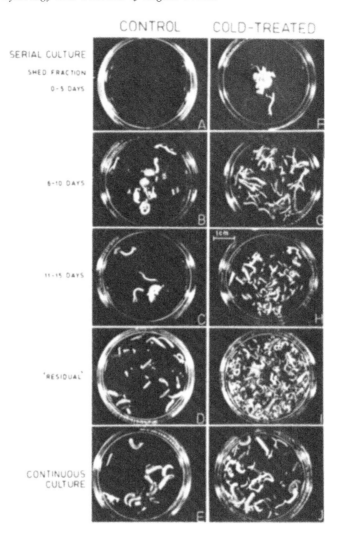

FIGURE 3.   Embryo formation on anther culture of *Datura innoxia*, with or without cold treatment, on liquid medium either as serial culture or continuous culture. (From Tyagi, A. K., Rashid, A., and Maheshwari, S. C., *Protoplasma*, 99, 11, 1979. With permission.)

Of these, the first three modes of origin occur to a varying degree in many plants such as *Atropa*[194] and *Nicotiana*[195] (Figure 4). However, it is possible that a particular mode is more frequent than another. For instance, in *Datura innoxia*[215] and *N. sylvestris*[195] it is the first mode which is prevalent, whereas in *N. tabacum* cv. White Burley[241] the second mode is prevalent. Less frequent is the third mode. Only *Hyoscyamus niger* pollen plants are of generative cell origin,[180] mode 4.

The multicellular pollen represents a proembryonal stage, and from this, a pollen embryo is formed in stages which are reminiscent of zygotic embryogeny. This is clearly seen in those genera which respond to form haploids on a simple mineral-sucrose medium. Contrary to it, in many of the monocots and some dicot plants where hormones are required for the initiation of divisions in a pollen, the multicellular structure formed regresses to result in a callus. From this tissue, plantlets are possible either on hormone-free medium or a medium having reduced level of hormone. Pollen embryos or pollen calli are freely dispersed in an anther and emerge by piercing the anther wall. There is no specific mode of liberation and many such structures which fail to come out of anther remain arrested in development.

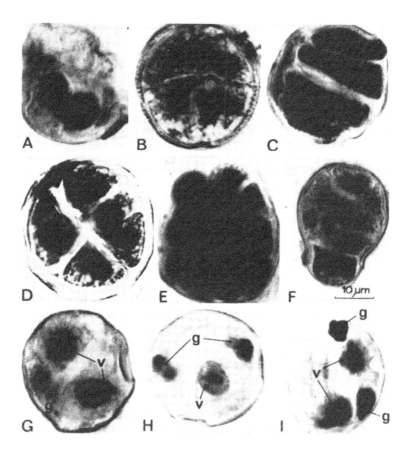

FIGURE 4. Stages in the development of pollen embryos of *Nicotiana tabacum*. (A) Microspore having formed two identical nuclei. (B, C) Four-celled pollen embryos. (D, E) Globular embryos with identical cells. (F) Pollen embryo with cells derived from the vegetative cell, and an undivided generative cell. (G, H) Pollen showing, respectively, division of vegetative (v) nucleus and generative (g) nucleus. (I) Pollen in which generative as well as vegetative cell have divided. (From Rashid, A. and Street, H. E., *Protoplasma*, 80, 323, 1974. With permission.)

Although in most of the genera the pollen embryos have been described to be freely dispersed, in one investigation on *Nicotiana* the pollen embryos are described[72] to be adhering to a supporting tissue which is either anther wall, connective, other embryos, or even clusters of dead pollen. Also, the position of rupture of pollen exine in a pollen embryo is described to be fixed so as to expose the plumular end of embryo. These aspects of pollen embryo formation need to be persued for confirmation.

F. Problem of Albino Pollen Plants in Cereals

The low frequency of formation of pollen plants in graminaceous genera is further accentuated by a high frequency of albinos which remain arrested in growth due to failure of differentiation of plastids into chloroplasts. To an extent, physiological factors effect albinism.

The frequency of albino plants depended on rice cultivar[21] as well as the temperature employed for culture of anthers.[30] The frequency of albino plants in wheat also increased with an increase in temperature of anther culture from 25 to 35°C.[7,266] As for cold treatment there are contradictory results. A prolonged cold treatment resulted in albino plants in rice, but not in barley. Anthers of rice containing uninucleate pollen grains when cold treated for 3 to 14 days at 10°C produced 90% green plants and 10% albinos, whereas an extended cold treatment of 21 days resulted in almost all albino plants.[62] However, in barley cv. Sabarlis an extended cold treatment did not result in a change of ratio of green to albino plants.[90]

The reasons for albinism are unknown. At one time, an albino plant was ascribed to originate from plastid-free generative cell, but albino plants have been shown to have plastids.[34,227] However, lack of 23S and 16S rRNA as well as fraction I protein in albino plants indicates that albinism is due to variation in plastid genome.[228] Other reasons advanced for albinism are that albino plants arise (1) from pollen with chromosomal aberrations[31] and (2) from pollen in which differentiation of plastids fails to occur following degeneration of plastids during meiosis (3) partly under genetic control.[80] In wheat incidence of albinism gradually decreased over a series of selfing.

A new finding[250] in this context, which needs to be extended, is the increased frequency of green plants in rice on treatment of callus tissue during differentiation with ABA (4 × 10^{-6} M).

VI. MICROSPORE CULTURE FOR INDUCTION OF HAPLOID

The culture of whole anthers does not permit experimentation with pollen as a free cell. By contrast, in pollen culture they can be plated, counted, and experimented with like microbes. Due to its haploid nature, pollen on mutation can increase the pool of genetic variability needed for crop improvement. Pollen, being haploid and a free cell, is reckoned to be a higher plant equivalent of a microbe. Higher yields of haploids are possible in pollen culture because the inhibitory influence of anther wall is lacking and the possible competition between developing embryos in the limitation of anther space is removed. In pollen culture one can unequivocally eliminate the contaminating proliferation from anther wall which is always in existence in anther culture. Also, in pollen culture the regulation of haploid formation can be studied with a degree of precision that is not possible in anther culture.

A. History of Pollen Culture

For pollen culture, a beginning was made with *Brassica*[105] when isolated mature pollen were reported to form cell aggregates on a mineral medium fortified with coconut milk, a medium earlier found to be suitable for anther culture of this plant. Isolated pollen of *Brassica* do form embryos, but the success is limited to the formation of one to two embryos from the entire population of pollen from an anther.[133] Another plant in which pollen culture was possible is *Petunia*.[9] Here too, the success was limited to the formation of a few cellular structures in a few cultures.

A simple approach adopted for the culture of isolated pollen of *Lycopersicon*[213] was to provide a nurse tissue. An intact anther cultured on salt agar medium fortified with auxin served as nurse tissue. About ten pollen were pipetted over a filter paper disc and this was placed over the anther. The success ranged from 0 to 60% of the pollen forming tissues in different cultures. The calli developed were haploid, but failed to differentiate into plantlets. By contrast, microspores of tobacco,[168] at the time of first mitosis, when placed directly over the petal callus of *Petunia* (serving as nurse tissue) formed embryos which developed into plantlets. These results indicated that isolated microspores could be stimulated to divide, but they required a special milieu which was provided by the nurse tissue.

The first claim of formation of embryos in cultures of isolated pollen was in *Datura*.[161] The success was ascribed to (1) induction of pollen towards embryogenesis and (2) providing the necessary nutrients to the induced units. For the induction of pollen towards embryo formation, young buds, at the time of first pollen mitosis, were subjected to a cold treatment at 3°C for 48 hr. The index of induction was described as a disturbance in polarity of pollen mitosis leading to the formation of two identical nuclei. Necessary nutrients to isolated pollen were provided in the form of a conditioned medium, which comprised mineral salts supplemented with anther extract from 1-week-old anthers, on culture, at the rate of five anthers per milliliter of medium. The anther extract was considered to be a substitute for an intact anther wherein pollen readily form plants. Interestingly, for microspores of *Datura* the conditioning of medium could be done by taking extract from *Nicotiana* anthers.[156] Also the conditioned medium could be replaced by a defined medium supplemented with high levels of glutamine (800 mg/ℓ), serine (100 mg/ℓ), and inositol (5 g/ℓ).

Unfortunately, these results on pollen culture of *Datura* could not be substantiated by other workers. In a reinvestigation,[232] half the cultures failed to form embryos and in responding ones, the frequency was very low, one to ten embryos per culture raised from pollen of four anthers. In another reinvestigation,[259] occasionally a few embryos appeared in some cultures.

B. Pollen Culture From Precultured Anther

In contrast to the direct culture of pollen, which had little success, pollen isolated from precultured anthers[157,159] of *Nicotiana* readily responded to form plantlet. These results have been confirmed by a number of workers.[94,200,232,275,276] The success in pollen culture on isolation from precultured anthers is ascribed to the critical phase for pollen plant formation which passes within the anther during its preculture.[137]

However, in a work on pollen culture from precultured anthers one has to be certain whether the cultures raised are comprised exclusively of pollen or also young embryos. This is seen in tobacco,[94] where after a preculture period of 8 days, without any prior cold treatment, the pollen suspension isolated was comprised of three- to five-celled embryos. Also, one has to consider (1) stage of pollen at the time of preculture, (2) anther pretreatment — cold treatment, (3) duration of cold treatment, and (4) duration of anther preculture.

The period of anther preculture required for pollen culture is dependent on cold treatment of anthers and the stage of pollen in an anther. In tobacco cultivar atropurpurea,[275] 4 days of preculture without cold treatment was sufficient for a good response provided anthers were at pollen mitosis or bicellular stage, whereas a preculture of 4 to 8 days was of no help at microspore stage. At uninucleate stage, 8 days of preculture were required. For White Burley cultivar[232] of tobacco 7 days of preculture were required without cold treatment. This could be reduced to 5 days if anthers before culture were given a cold treatment for 4 days. With a longer cold treatment, 8 days, a good response was possible just after 3 days of anther preculture.

Pollen cultures raised from precultured anthers have also been a success in *Secale*,[272] *Hyoscyamus*,[275] *N. sylvestris*,[40] *Datura*,[232,259] *Solanum* spp.,[268] and *Saccharum*.[84] Of these, the investigations which deserve special mention are those of *Secale*,[272] *Nicotiana*,[276] and *Saccharum*.[84] In *Secale* an attempt was made to select pollen which are capable of division. Pollen from precultured anthers were layered on top of 30% sucrose and its centrifugation at 1200 *g* for 5 min resulted in some microspores forming a band at the top and the mature ones settling at the base. Pollen from the upper layer showed some divisions in a few cultures. However, in *Nicotiana* it was possible to have an enrichment of microspores capable of division on isolation from anthers precultured for 4 days. The selection was done on density gradient centrifugation, employing 0.25 *M* sucrose in Percoll stepwise diluted, at 150 *g* for 15 min. Among the pollen in interphase of Percoll, about 8% were capable of division. By making finer gradients and extending the period of preculture of anthers to 7 days, a fraction was obtained from which about 30% of the pollen divided. In microspores from precultured anthers of *Saccharum spontaneum* the frequency of units forming multicellular aggregates could be increased on removing nonviable microspores.[84]

A pollen culture from precultured anthers is more suitable for an enhanced response, in terms of pollen plants, than the anther culture. However, it is not suitable for any other study such as raising of mutants because it is not certain whether or not pollen divide or start to divide in anthers during the period of preculture.

C. *Ab Initio* Culture of Pollen

Success in *ab initio* culture of pollen[189] of *Nicotiana* leading to the formation of pollen plants was possible due to (1) realization that pollen capable of embryogenesis are different from those destined to form gametes and (2) adopting a new technique for pollen culture, in which (a) donor plants were grown at low temperature, (b) young buds from these plants were given a cold treatment, and (c) pollen from these buds were separated into embryogenic and nonembryogenic fractions. When young buds from plants of *N. tabacum* cv. Badischer Burley induced to flower at 18°C were subjected to an additional cold treatment at 10°C for 10 days, differentiation of pollen into two types could be seen. In a low frequency were small pollen that stained lightly with acetocarmine and the rest were large densely staining pollen. The former were considered to be embryogenic and the latter were nonembryogenic or gametophytic. The embryogenic nature of these smaller pollen could be confirmed by their separation from the rest of pollen. On centrifugation on low sugar-Percoll solution, the smaller pollen formed a band at the top and the bigger pollen pelleted at the base. These smaller pollen on culture formed embryos at a low frequency, 2% of the cultured pollen.

Further experiments[190] on pollen culture revealed that it is possible to isolate potentially embryogenic pollen from buds originating from plants flowering at 24°C as well as from plants flowering at 18°C, provided that in both cases the buds were given a cold treatment at 10°C prior to isolation of pollen. The buds from 24°C plants required at least 10 days of cold treatment, whereas buds from 18°C plants required just 7 days of cold treatment. Also, the frequency of embryo formation was much higher (up to 5% of cultured pollen) in pollen cultures originating from plants flowering at 18°C than in pollen from plants flowering at 24°C (1% or less). This indicated that buds from plants induced to flower at low temperature had a higher frequency of pollen that were preconditioned towards embryo formation.

That the pollen can be preconditioned towards embryo formation on the plants[191] was substantiated by further experiments. It was possible to have higher frequencies of embryo formation, 11 to 27% of the cultured pollen in different experiments, provided the pollen were taken from plants induced to flower at a still lower temperature (15°C) than the 18°C employed earlier. The buds were given a cold treatment at 10°C for 10 days. A consistent high frequency of embryo formation (Figure 5). 30% of the cultured pollen,[193] was possible when the buds were taken from plants in the early period of flowering. Also, the flower-buds taken for cold treatment were slightly older than employed earlier. At the older stage of buds (petal length 2.5 cm) the gamete-forming pollen were at an advanced stage of development and a better separation of two types of pollen was possible on centrifugation in Percoll at 15°C. About 15% of the pollen was comprised of the embryogenic fraction. By contrast, only 3 to 5% of the pollen formed the embryogenic fraction if the buds were taken from plants flowering at 24°C. This clearly indicated that low temperature increased the frequency of embryogenic pollen.

The results reported above were possible due to increasing the frequency of dimorphic pollen and selection of embryogenic pollen. However, the high frequency of pollen dimorphism occurring in nature is on record in peony and in this plant callus formation is possible by 2 to 3% of the pollen without any selection.[166] Also in *Nicotiana tabacum*[64] and *N. rustica*[96] embryo formation is possible, on direct culture of pollen. For this, pollen are first cultured in water for a few days and then the nutrients are supplied. A preculture of pollen in pure water is described to be essential for the success. During this period, described as starvation, some physiological changes responsible for embryogenesis seem to take place.[124] Similarly, a preculture of barley[270] pollen in 0.3 *M* mannitol solution for 5 to 7 days, and then reculture in nutrient medium having 6% sucrose, resulted in the formation of embryos.

FIGURE 5. *Ab initio* culture of pollen of *Nicotiana tabacum* cv. Badischer Burley showing early stages of embryos developing from pollen grains. (From Rashid, A. and Reinert, J., *Protoplasma*, 116, 155, 1983. With permission.)

On direct culture of total pollen, embryo formation occurs at a very low frequency. However, high frequency embryo formation is possible in tobacco if two-stage fractionation of pollen, employing Percoll, is done. Fractionations are done before culture of pollen in water as well as after water culture. In this way it is possible to have 40% of dividing pollen. These results are consistent with the earlier observations[189,190] that embryogenic pollen are present on the plant and their determination to embryo formation can be stimulated by cultural conditions which are cold treatment[189,190] and starvation.[124] However, a relationship, if any, between cold treatment and starvation remains to be resolved.

D. Nutritional Requirement of Isolated Pollen

Pollen isolated from precultured anthers[157] of *Nicotiana* were cultured on mineral medium supplemented with glutamine, serine, and inositol. Subsequent workers employed either the same medium[42,78,200] or only glutamine[236,275,276] implying thereby that glutamine alone is essential. Also, for pollen of *Solanum* spp. glutamine was found to be essential, but in addition, asparagine was required. The medium with glutamine and asparagine was sufficient for *S. tarjense*, but was insufficient for *S. tuberosum* var. Pentland Crown and had to be supplemented with auxin, NAA.[268] In contrast to these reports, pollen from precultured anthers of *Datura* readily responded to form embryos on simple mineral-sucrose medium.[232,259]

In the initial attempt at *ab initio* pollen culture of *Nicotiana*, glutamine and asparagine were employed.[189] However, soon it was realized that nutritional requirements of isolated embryogenic pollen are surprisingly simple. The pollen are able to initiate and complete embryogenesis on a simple mineral-sucrose medium[190] provided the pH of the medium is near neutral, 6.8. It appears that high pH has a stabilizing role on pollen in isolation. These results have been confirmed in pollen culture of *N. rustica*.[124]

VII. PHYSIOLOGY AND BIOCHEMISTRY OF POLLEN PLANT FORMATION

Although a large number of factors are known to affect formation of pollen plants, an understanding of the process of induction, how a pollen can be diverted towards a sporophytic mode instead of gametophytic one, is far from complete. This basic information will not only be helpful for achieving higher induction frequencies, but will also help in the induction of haploids in recalcitrant types.

A. The Induction Process
On the basis of available information the process of induction of pollen towards sporophytic mode is described as occurring in two divergent ways.

1. Induction of Embryogenic Pollen in Response to Culture of Anthers
To begin with, induction in *Nicotiana* was ascribed in response to culture of anthers.[241] Following induction, 6 to 12 days of anther culture, two types of pollen could be recognized. Densely staining pollen were described as nonembryogenic and lightly staining pollen as embryogenic. Cytochemically in nonembryogenic pollen there was a four- to sixfold increase in RNA and protein, whereas in embryogenic pollen there was loss of RNA and protein.[8] At the subcellular level,[49,50] in a pollen destined to be embryogenic, during the induction process appeared multivesiculate bodies resembling lysosomes which were ascribed to be responsible for the breakdown of cytoplasm. Much of the gametophytic cytoplasm present initially in the vegetative cell was degraded and by the end of the induction process (12 days of culture), in terms of cell organelles, only a few mitochondria and structurally simplified plastids were left in the vegetative cell while ribosomes were almost completely eliminated. However, in a study of induction process in *Datura*[51] these changes were not recorded, casting doubts about the contention of induction processes in tobacco.

Also, in *Hyoscyamus niger* the induction of embryogenic pollen grains is described to occur on culture of anthers[182-184] at the uninucleate stage of pollen. The embryogenically determined grains are confined to the periphery of anther locule. Differentiation of embryogenic pollen is not an abrupt process, but a gradual cessation of gametophyte differentiation. This is supported by the observations that the embryogenic pollen grain continued the basic gametophytic programs, such as vacuolation and asymmetric division, to form generative and vegetative cells. During this period certain cytological features[201] of these pollen indicate a program of redifferentiation leading to continued growth and development. The main cytological features recorded were (1) an increased ratio of dispersed to condensed chromatin in the nucleoplasm, whereas nonembryogenic pollen in vivo and in vitro possessed a low ratio of dispersed to condensed chromatin. This is consistent with the observations in several plant and animal systems where cells differentiating into new developmental pathways have been shown to have decondensation and dispersion of chromatin; (2) following first haploid

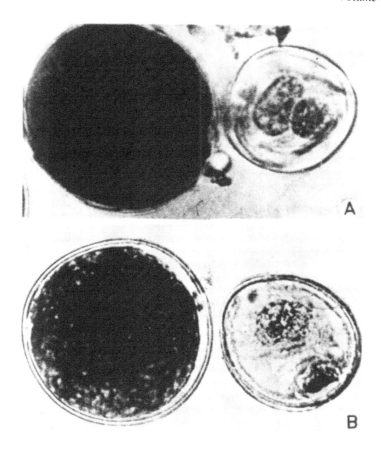

FIGURE 6.   Pollen dimorphism in *Nicotiana tabacum* cv. Badischer Burley (A) and *Hordeum vulgare* cv. Sabarlis (B). Large densely staining pollen are gametophytic and small lightly staining pollen are embryogenic. (From (A) Heberle-Bors, E. and Reinert, J., *Naturwissenschaften*, 67, 311, 1980, and (B) Dale, P. J., *Planta*, 127, 213, 1975. With permissions.)

mitosis, the generative cell of embryogenic cell maintained its large granular nucleolus and high ratio of dispersed to condensed chromatin; (3) the generative cell of potentially embryogenic pollen differed from the generative cell of nonembryogenic pollen in having increased area and complexity of cytoplasmic membranes, increased mitochondrial volume, and the presence of plastids at all stages of development. These changes are considered to be essential for continued development into an embryo. In this plant, embryo arises from divisions of generative cell.

Also of interest in this context is the recent[252] identification of pollen capable of forming haploids in anthers of rice on the basis of cytological characters. About 4% of the microspores which accumulate abundant amorphous lipid in the first few days of culture are described as androgenic. This finding needs to be persued.

2. Pollen Dimorphism in Relation to Pollen Plant Formation

In contrast to the induction process described as occurring in response to culture and resulting in differentiation of two types of pollen,[241] embryo-forming pollen were demonstrated in anthers of barley[37] and tobacco[86] at the time of culture. The embryogenic pollen are small with a cytoplasm that stains less intensely with acetocarmine than the gametophytic pollen (Figure 6). On culture of anthers, embryos arise also from smaller pollen in wheat.[244,285] An unequivocal evidence for the embryogenic nature of these smaller pollen in tobacco is their selective isolation and development into embryos.[189-193]

a. Induction of Embryogenic Pollen — Role of Cold Treatment

Of the many factors affecting pollen plant formation in anther culture, the most significant is cold treatment to buds or anthers prior to culture. How cold treatment affects this response remains to the resolved. Except for causing a delay in senescence of anthers, other conclusions based on experimental results of cold treatment on anther culture are inconsistent and suggested roles of cold treatment are speculative. Nonetheless, an inquiry into the principal factor contributing towards the success of experiments on *ab initio* culture of pollen of *Nicotiana*[189-193] led to the conclusion that cold treatment brings about the differentiation of embryogenic pollen. On making tobacco plants flower at a low temperature, 18°C and 15°C, there was a corresponding increase in the frequency of smaller embryo-forming pollen. This became more apparent when young buds were given an extended cold treatment at 10°C for 10 days. During cold treatment of buds a small number of pollen remained small in size; each of them retained a clear cytoplasm and differentiated into embryogenic pollen. This was in contrast to the majority of pollen which acquired granular cytoplasm and differentiated into gametophytic pollen.

B. Differentiation of Embryogenic Pollen — Two-Step Process

The role of cold treatment in the differentiation of embryogenic pollen could be further substantiated when pollen cultures were raised from buds which were not given cold treatment. For this, older buds had to be employed because from them it is possible to detect and separate two types of pollen. When pollen cultures were raised from donor plants induced to flower at 18°C, the embryos failed to appear, but embryo formation could be induced if the cultures were given a cold treatment at 10°C for 10 days, whereas if cultures were raised from donor plants flowering at 15°C, the embryos appeared directly, but at a low frequency which could be increased by giving the cultures a cold treatment.[192] From these experiments it can be concluded that differentiation of embryo-forming pollen is a two-step process; it initiates on the plant and is completed in culture. Differentiation on the plant results in pollen dimorphism, and differentiation in culture leads to the formation of embryos. Differentiation of embryogenic pollen is a gradual process.

Even in relatively less responsive species, like *N. sylvestris*,[187] induction of embryos in *ab initio* pollen cultures was possible by an application of cold treatment to plants, buds, and pollen cultures. Young buds were taken from plants induced to flower at 15°C, and these buds were subjected to cold treatment at 10°C. During cold treatment of buds differentiation of embryogenic pollen takes place. However, in *N. sylvestris* this differentiation remained incomplete, even after 15 days of cold treatment at 10°C. To achieve a complete differentiation, an additional cold treatment to cultures for 12 days at 10°C resulted in differentiation of embryos.

It remains to be resolved as to how cold treatment brings about the differentiation of embryo-forming pollen in *Nicotiana*.[188] A subcellular study of embryogenic pollen of *Nicotiana* from plants induced to flower at 15°C is quite revealing. Cold treatment possibly brings about the differentiation of embryo-forming pollen by the repression of gametophytic differentiation.[197] The gametophytic pollen has a thick exine, a dense cytoplasm which is enriched with ribosomes, and has well-developed mitochondria. Such a pollen can germinate, even in the presence of metabolic inhibitors like actinomycin-D and cycloheximide, thereby indicating that it has preformed messengers. By contrast, these features of gametophytic differentiation are lacking in embryogenic pollen (Figure 7). The embryogenic pollen has a thin exine, an attenuated cytoplasm that is poor in ribosome, and has condensed mitochondria. Such a pollen neither germinates nor forms embryo in the presence of metabolic inhibitors.[192] Possibly because of cytoplasmic attenuation, the embryogenic pollen remain small in size. Also, the condensation of mitochondria may impair their ability to germinate and help these pollen turn towards the sporophytic pathway. A parallel can be drawn between mitochondrial

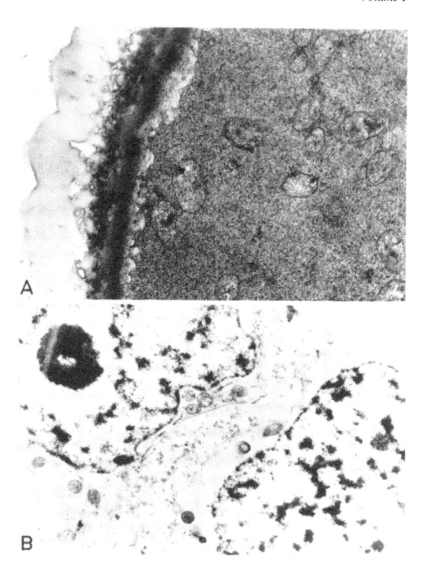

FIGURE 7. Ultrastructure of gametophytic and embryogenic pollen of *Nicotiana tabacum* cv. Badischer Burley. (A) Gametophytic pollen showing dense cytoplasm and well-developed mitochondria. (B) Embryogenic pollen, having undergone symmetric division showing sparse cytoplasm and condensed mitochondria. (B from Rashid, A., Siddiqui, A. W., and Reinert, J., *Protoplasma*, 107, 375, 1981. With permission.)

condensation and embryogenic potential and mitochondrial attenuation and cytoplasmic male sterility.[129,131] The other effects of cold treatment are the disturbance of cytoplasmic distribution and nuclear division. These features are considered to be conducive to embryo formation, but do not seem to be essential. Cold treatment also increases the viability of embryogenic pollen by arresting the process of degeneration. However, this is a secondary effect.

Therefore, repression of gametophytic differentiation[188,197] is probably the primary event in the process of differentiation of embryogenic pollen of *Nicotiana*. One can arrive at this conclusion if pollen dimorphism is seen on the plants. Alternately, if embryo-forming pollen are seen in response to culture of anthers, as in *Hyocyamus*, their differentiation is a result of gradual cessation of gametophytic differentiation.[201] In both these genera, although embryogenic pollen undergo asymmetric division, embryo is formed by the division of vegetative cell in *Nicotiana* and by the division of generative cell in *Hyoscyamus*. However, in both these genera a small fraction of embryogenic pollen undergo symmetric division, and both the cells formed participate in the formation of an embryo. This type of pollen in *Hyocyamus*[202] does not show any carryover of gametophytic influence following embryogenic induction.

C. Formation of Pollen Embryos — An Impact of Culture

The embryogenic pollen grains of *Nicotiana*,[197] in their features of sparse cytoplasm, paucity of ribosomes, condensed mitochondria, and inability to germinate, present the profile of a cell that is metabolically inactive. In these features the embryogenic pollen are comparable to unfertilized eggs of many angiosperms[100] and this similarity may be taken as an evidence of their sporophytic potential.

Activation of embryogenic pollen of *Nicotiana* in vitro, the second part of their differentiation, which results in the formation of embryos, is possible on simple mineral-sucrose medium,[190-192] provided the pH is near neutrality. However, before the initiation of division, the most significant aspect of pollen embryo formation is the appearance of a fibrillar wall around the pollen cytoplasm.[198] A thin fibrillar wall appears to be the unique feature of this process and it possibly helps the pollen to embark on embryogenesis by providing a check on efflux and influx of substances between pollen and its new environment — the culture. Alternately, the wall may help the pollen acquire polarity. Another significant change is the redifferentiation of mitochondria, which reflects the change of pollen from an inactive to an active state.

D. Biochemistry of Pollen Embryogenesis

A beginning towards biochemical aspects of pollen plant formation has been possible in *Hyoscyamus niger*. In this system, on anther culture, it is possible to identify the embryogenic pollen[182,183,185] on monitoring RNA synthesis. In nonembryogenic pollen, only in vegetative cell is RNA synthesis detectable, whereas in embryogenic pollen, generative as well as vegetative cells show RNA syntheses.[182] Also, in this system pollen grains become committed to form embryos within hours of culture. This is also inferred by detection of RNA synthesis.[183] Furthermore, it is possible to identify embryogenic pollen, even at the uninucleate stage. On culture of anthers, the majority of pollen grain incorporate labeled amino acids (arginine, leucine, lysine, and tryptophan). However, protein synthesis persists only in a small number of pollen which are uninucleate, nonvacuolate, and densly staining, embryogenically determined pollen grain, confined to the periphery of anther locule.[185]

The RNA synthesized in embryogenic pollen grains is polyadenylated (carries the information for embryogenic division) and is evidenced by its binding to poly-U.[184] Only in embryologically determined grains is this binding seen, while the majority of grains, which are nonembryogenic, do not show binding. Also, in embryogenic grains this binding does not occur in the presence of actinomycin-D. This indicates that poly-A containing mRNA is newly synthesized by certain pollen grains as they establish contact with the medium and embark upon embryogenesis. Further, that mRNA is probably concerned with embryogenesis becomes clear from the result that uninucleate pollen grains do not bind ^{3}H-poly-U at the time of culture or at earlier stages of their ontogeny.

VIII. REGENERATION CAPACITY OF POLLEN PLANT/CELL

Pollen plants/embryos developing from pollen grains of *Datura*[61] and other plants are described as forming secondary embryos on hypocotyl. This indicates the high regenerative capacity of haploid plants. It is also evidenced in cell suspension cultures originating from

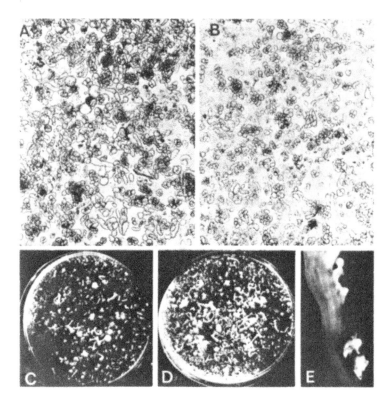

FIGURE 8. Profiles of diploid and haploid cells of *Atropa belladonna* and their regeneration potential. (A) Diploid cells. (B) Haploid cells; note smaller size of haploid cells. (C, D) Formation of embryos from suspension cultures of diploid and haploid cells, respectively. Note higher regeneration potential of haploid cells. (E) Plantlet developed from haploid embryo showing the formation of secondary embryos on hypocotyl. (From Rashid, A. and Street, H. E., *Plant Sci. Lett.*, 2, 89, 1974. With permission.)

a pollen plant of *Atropa*.[196] Haploid embryoids differentiated from cell suspension formed secondary embryos on hypocotyl. A comparison of haploid and diploid cells of *Atropa* (Figure 8) revealed that haploid cells are small in sizes, fast-growing, and regenerated into embryos at a higher frequency than diploid cells. The fast growth of haploid cells can be attributed to their small size, but a reason for higher regeneration potential of haploid cells and embryos remains to be understood. It is also quite paradoxical that tissues derived from pollen readily regenerate in wheat, barley, and rice, whereas regeneration of plants from diploid tissues of somatic cell origin in these plants is quite infrequent. Also, in *Datura innoxia*[219] haploid callus regenerated far more readily than diploid callus. However, no difference was found in haploid and diploid tissues of this plant by other workers.[211] This controversy was resolved when it was found that shoot initiation[58] in haploid callus precedes that in diploid.

Since diploid tissues progressively lose their capacity for regeneration and become polyploid and aneuploid, it remains to be seen how long haploid tissues remain haploid and totipotent. In *Atropa*,[196] a haploid cell line showed a decline in proportion of haploid cells and loss of embryogenic potential. An increase in concentration of cytokinin resulted in a rapid departure from haploidy. Contrary to this, a pollen callus line of *Zea* retained its haploidy and regeneration potential. More than 20,000 haploid plantlets were obtained from this line during 2 years.[279]

IX. PLOIDY OF POLLEN PLANT

The majority of pollen plants derived from anther culture of tobacco were found to be haploid.[14,229] However, it is not uncommon to find diploid, polyploid, mixoploid, and aneuploid plants on anther culture. For a fast and accurate determination of ploidy of pollen plants, a flow cytometric method can be adopted.[212]

The origin of nonhaploid plants can be viewed in terms of nutritional requirements of anthers to form pollen plants. In monocots in general, and cereals in particular, where growth hormones are an obligate requirement and the origin of pollen plants is through callus formation, irregular mitoses of pollen callus are ascribed to result in aneuploid and mixoploid plants.[89,147] Also, in monocots the origin of plants from somatic tissues should not be ruled out.[154] Therefore, in view of the deleterious effects of hormones, their utility for increasing the response should be discounted in genera which readily form pollen plants on a simple medium. However, even in this category of plants the ploidy of pollen plants formed is dependent upon the stage of pollen at the time of anther culture and growth conditions of donor plant. That the diploid plants are microspore-derived is evidenced from their genetic analysis. F_1 progenies from them are very uniform in rice[155] and wheat.[167]

The increase in ploidy of pollen plants is ascribed to endomitosis and nuclear fusion in pollen during culture of anthers.[55,153] The possibility of nuclear fusion has been proposed in *Datura*.[235] In some grains division may be simultaneous in vegetative and generative cells. However, in generative cells there may be two successive cycles of DNA synthesis, and it may be limited to one in vegetative nucleus. A 3n plant is possible if division on a common spindle occurs by diplochromosomes of endoreduplicated generative nucleus and those of the vegetative nucleus. Also, a study of haploid, diploid, and triploid plant formation in *Petunia*[71] has indicated that nonhaploid embryoids developed due to nuclear fusion at an early stage of pollen embryogenesis.

A. Diplodization of Haploid Cell/Plant

Haploid plants are sterile and fail to set seed. The pollen produced by them are also nonfunctional. For want of homologous chromosomes there is no pairing and there is abnormal meiosis. Therefore, diploid plants are to be produced for their utility as pure lines in plant breeding.

Diplodization of haploid cells is possible by application of colchicine. Haploid cells of *Atropa*[196] could be diplodized by adding 1% colchicine to exponentially growing suspension culture for 24 hr. However, a simpler method is spontaneous diplodization of cells during culture. Segments of petiole,[119,196] stem, root, and leaf[111] are cultured on suitable medium for callus formation. The frequency of diploid cells increases with an increase of culture duration. The process is hastened in *Atropa*[196] by an increase in the level of cytokinin.

Individual pollen plants, which undergo spontaneous doubling[229] at a very low frequency, can be diplodized while young by immersing them in solution of colchicine for 2 to 3 days. Alternately, colchicine is applied to axillary buds and the main plant is decapitated. The sprouted axillary plants are diploid. Besides increasing the number of chromosomes, colchicine has other harmful effects such as chromosome and gene abnormalities.[17] Therefore, dimethylsulfoxide (DMSO) is employed along with colchicine for increasing the efficiency of chromosome doubling.[112] More than 50% of haploid plants of *N. rustica*[27] could be diplodized on treatment of 0.4% colchicine and 2% DMSO for 5 hr.

X. HAPLOID PLANT IN PLANT BREEDING

The utility of haploid plants in plant breeding is dependent on the nature of crop, whether it is self- or cross-pollinated, or diploid or autopolyploid.

A. Self-Pollinated Crops

The crops are either diploid or allopolyploid. An allopolyploid is a functional diploid and is often referred to as amphidiploid. The advantages of employing dihaploid lines in plant breeding are as follows:

1. A dihaploid line is the fastest possible way of obtaining a homozygous line — in one generation, whereas the conventional method of repeated-selfing requires six to seven generations.
2. Dihaploid lines can be reliably tested at any stage of the breeding program.
3. Dihaploid lines exhibit additive variance only, whereas early-selfed generations contain a considerable proportion of nonselectable, nonadditive variance. Therefore, employing dihaploid lines it is possible to have a reliable selection in early generations and an earlier and more reliable assessment of the breeding value of hybrid populations.[199]
4. Dihaploid lines have the potential for improving the efficiency of existing breeding method. In a recurrent selection method, consisting of cycles of intensive intercrossing and selfing of selected genotypes, selfing may be replaced by dihaploid production because an assessment of dihaploid lines is much more reliable than assessing the selfed progeny of heterozygous plants.
5. Homozygous lines produced by repeated selfing are adapted to climatic conditions of the region where they are grown, whereas haploid lines have not undergone any selection. This may be advantageous for producing varieties for regions with different climatic conditions.[109]

B. Cross-Pollinated Crops

The application of haploidy in cross-pollinated diploid crops is also based on the use of dihaploid lines. However, owing to inbreeding depression, these lines cannot be used directly, but only as parental inbred lines. The advantages of dihaploid inbred lines are as follows:

1. When inbred lines are developed through haploids, all barriers to repeated selfing, such as dioecy, self-incompatibility, and long juvenile period, are by-passed. Doubled haploids and their F_1 hybrids regenerated via anther culture from self-incompatible *Solanum chacoense*[20] were self-compatible.
2. An additional advantage for self-incompatible crops is that selection for weak incompatibility alleles or genetic background inhibiting the activities of such alleles is avoided.
3. The time saving is particularly apparent in biennial crops and in crops with long juvenile periods. Only through haploidy can inbred lines be produced in these plants.
4. For developing all male hybrid varieties in a dioecious crop, super male lines are needed which can only be obtained through haploidy from normal male plants. For instance, male asparagus hybrids are generally superior to female ones.
5. A large variation in vigor is expected in dihaploid lines. Selection of lines with a relatively high vigor may be carried out so that an economical production of single cross-seeds is possible. Weak inbred lines may also be a valuable genotype.

C. Autopolyploid Crops

It is difficult to foresee the practical application of haploidy in autohexaploid crops, because halving of their chromosome number will lead to sterile trihaploids. However, in autotetraploid crops two successive cycles of haploidization are possible. This has in fact been realized in potato.

Possible applications of dihaploids in plant breeding are (1) simplified genetics and breeding at diploid level and (2) improved crossability with diploid wild species, which are indispensable in crops like potato. It is possible to have germplasm enhancement with potato haploids. Wild species of potato contain valuable characters and the allelic diversity necessary to increase heterozygosity, but they do not tuberize. Therefore, hybridization with dihaploids is a valuable method of putting this germplasm into a form that tuberizes. Hybrid species were produced by crosses between 40 *Solanum tuberosum* (group tuberosum) haploids (2n = 2× = 24) as females with wild 2n = 2× = 24 *Solanum* spp., *S. berthaultii*, *S. boliviense, S. canasense* and many others as males. Several important results were obtained; in particular, variation in tuberization was striking and discrete.[82]

However, there are difficulties in working with dihaploids because of the prevalence of male sterility, lowered female fertility, and poor flowering. In addition, introduction of self-incompatibility may hamper self-fertilization and restrict crossability. A basic limitation of dihaploids is their restriction to two alleles per locus. For this reason diploid potatoes cannot be expected to compete with tetraploids.

Monohaploid from an autotetraploid crop is the only tool for the production of homozygous tetraploid of this crop such as potato.

D. Pollen Plant as Natural Variant

Since each pollen is the product of meiosis, it possesses a unique genome, as a result of pairing and recombination, and it is very likely that pollen plants are readily available, desired variants. Further, in them the process of selection is greatly simplified due to the greater homozygosity of doubled haploid lines, which is greater than in the diploid inbred lines.[24] One expects a higher efficiency of selection among dihaploid lines than in diploid breeding. Because of these advantages, new cultivars have been released efficiently from a smaller population of pollen-derived plants. In tobacco,[152] the variety "F-211" is more resistant to bacterial wilt and has mild smoking quality, and high-yielders have been obtained in rice[282] and wheat.[253]

In rice, F_1 hybrid between subspecies *japonica* and *indica* is partially sterile and progeny segregates vigorously. However, the segregation is controlled on anther culture. Pollen-derived plants are homozygous, and 50% of them were wholly fertile. A new cultivar has been selected from fertile lines and released for cultivation in China.[88] As for barley, within 2 years of trial, pollen-derived "Mingo" was the highest yielding barley in Ontario in 1980.[85,110] Selection within 2 years was possible due to dihaploid breeding, wherein one can evaluate four times more lines because the evalution is possible on plot basis rather than individual plant basis, as done in diploid breeding. A new short-statured rice from anther callus of "Calrose-76" has been obtained at the USDA.[209] It may be a potential new source of dwarfing.

XI. UNEXPLORED POTENTIALS OF POLLEN PLANT FOR CROP IMPROVEMENT

Besides their prime utility as "pure line" in crop breeding, and their utility in mutational breeding via haploid cells, pollen plants are useful for new chromosomal and cytoplasmic variations.

Dihaploids can serve for transfer of genes from one species to another. If an interspecific hybrid in which the chromosomes showing partial pairing in meiosis is employed for anther culture, it is very likely that some homozygous lines with heterozygous translocation may result. In such cases, genes for higher yield, good quality, or disease resistance may be transferred from one species to another.

Haploids could be a source of new chromosomal combination in distant hybrids. When *Triticum aestivum* is crossed with *Secale cereale*, the F_1 hybrid contains 21 univalent wheat chromosomes and 7 univalent rye chromosomes. Owing to random redistribution of uni-valents during meiosis the microspores of hybrid can have various chromosome combinations.

Pollen plants could be a potential source of cytoplasmic variation. Since cytoplasm of a microspore does not participate in normal fertilization, it might not be conserved genetically, and is likely to be a new source of variation.

XII. MICROSPORE AS HIGHER PLANT-EQUIVALENT OF MICROBE

Microspores are reckoned to be a higher plant-equivalent of the microbe.[189,191] This is supported by the success in induction of pollen plants from the culture of isolated pollen. However, this possiblity is to be examined in light of the limitations of the pollen culture technique and its efficiency, compared with the alternate system, haploid protoplast.

At present, only in a few systems, tobacco,[191] datura,[234] peony,[166] and barley,[270] has it been possible to have pollen plants from isolated pollen grains. Of these, tobacco is not a true microspore system suitable to obtain solid mutants. To an extent, datura can be described to be a microspore system, but in this plant frequency of embryo formation by isolated pollen grain is very low and inconsistent. These are the basic limitations of these systems. Therefore, at this stage we are far from the objective, the microspore as a higher plant-equivalent of the microbe. Continued efforts in this direction are likely to be rewarding.

REFERENCES

1. **Amos, J. A. and Scholl, R. L.,** Induction of haploid callus from anthers of four species of *Arabidospsis*, *Z. Pflanzenphysiol.*, 90, 33, 1978.
2. **Anagnostakis, S. L.,** Haploid plants from anthers of tobacco. Enhancement with charcoal, *Planta*, 115, 281, 1974.
3. **Babbar, S. B. and Gupta, S. C.,** Promotory effect of polyvinyl-pyrrolidone and L-cysteine HCl on pollen plant production in anther cultures of *Datura metel*, *Z. Pflanzenphysiol.*, 106, 459, 1982.
4. **Barclay, I. R.,** High frequencies of haploid production in wheat *Triticum aestivum* by chromosome elimination, *Nature (London)* 265, 410, 1975.
5. **Bennett, M. D. and Hughes, W. G.,** Additional mitosis in wheat pollen induced by etherel, *Nature (London)*, 240, 566, 1972.
6. **Bennet, M. D., Finch, R. A., and Barclay, I. R.,** The time, rate and mechanism of chromosome elimination in *Hordeum* hybrids, *Chromosoma*, 54, 175, 1976.
7. **Bernard, S.,** In vitro androgenesis in hexaploid triticale — determination of physical conditions increasing embryoid and green plant production, *Z. Pflanzenphysiol.*, 85, 308, 1980.
8. **Bhojwani, S. S., Dunwell, J. M., and Sunderland, N.,** Nucleic acid and protein contents of embryogenic pollen, *J. Exp. Bot.*, 24, 863, 1973.
9. **Binding, H.,** Nuclear and cell divisions in isolated pollen of *Petunia hybrida* in agar and suspension cultures, *Nature (London) New Biol.*, 237, 283, 1972.
10. **Bingham, E. T.,** Haploids from cultivated alfalfa *Medicago sativa*, *Nature (London)*, 221, 865, 1969.
11. **Blakeslee, A. F., Belling, J., Farnham, M. E., and Bergner, A. D.,** A haploid mutant in the Jimson weed, *Datura stramonium*, *Science*, 55, 646, 1922.
12. **Blaydes, D. F.,** Interaction of kinetin and various inhibitors in the growth of soybean tissue, *Physiol. Plant*, 19, 74, 1966.
13. **Bonga, J. M. and McInnis, A. H.,** Stimulation of callus development from immature pollen of *Pinus resinosa* by centrifugation, *Plant Sci. Lett.*, 4, 199, 1975.
14. **Bourgin, J. P. and Nitsch, J. P.,** Obtention de *Nicotiana* haploides a partir d'etamines cultivees in vitro, *Ann. Physiol. Veg.*, 9, 377, 1967.
15. **Brewbaker, J. L.,** Pollen cytology and self-incompatibility systems in plants, *J. Hered.*, 48, 271, 1957.
16. **Bui-Dang Ha, D. and Pernes, J.,** Androgenesis in pearl millet. I. Analysis of plants obtained from microspore culture, *Z. Pflanzenphysiol.*, 108, 317, 1982.
17. **Burk, L. G.,** Green and light yellow haploid seedlings from anthers of sulpher tobacco, *J. Hered.*, 61, 279, 1970.
18. **Burk, L. G., Gerstel, D. U., and Wernsman, E. A.,** Maternal haploids of *Nicotiana tabacum* from seed, *Science*, 206, 585, 1979.

19. **Cappadocia, M., Cheng, D. S. K., and Ludlum-Simonette, R.,** Plant regeneration from in vitro culture of anthers of *Solanum chacoense* and interspecific diploid hybrids *S. tuberosum* × *S. chacoense, Theor. Appl. Genet.,* 69, 139, 1984.

20. **Cappadocia, M., Cheng, D. S. K., and Ludlum-Simonette, R.,** Self-compatibility in doubled haploids and their F₁ hybrids, regenerated via anther culture in self-incompatible *Solanum chacoense, Theor. Appl. Genet.,* 72, 66, 1986.

21. **Chaleef, R. S. and Stolarz, A.,** Factors influencing the frequency of callus formation in cultured rice anthers, *Physiol. Plant.,* 51, 201, 1981.

22. **Chase, S. S.,** Monoploid frequencies in a commercial doublecross hybrid maize and in its component singlecross hybrids and inbred lines, *Genetics (Princeton),* 34, 328, 1949.

23. **Chase, S. S.,** Androgenesis — its use for transfer of maize cytoplasm, *J. Hered.,* 54, 152, 1963.

24. **Chase, S. S.,** Monoploids and monoploid-derivatives of maize *(Zea mays), Bot. Rev.,* 35, 117, 1969.

25. **Chaturvedi, H. C. and Sharma, A. K.,** Androgenesis in *Citrus aurantifolia, Planta,* 165, 142, 1985.

26. **Chien, Y. C. and Kao, K. N.,** Effects of osmolality, cytokinin and organic acids on pollen callus formation in *Triticale* anthers, *Can. J. Bot.,* 61, 639, 1983.

27. **Chowdhury, M. K. U.,** An improved method for dihaploid production in *Nicotiana rustica* through anther culture, *Theor. Appl. Genet.,* 69, 199, 1984.

28. **Chowdhury, M. K. U.,** A comparative study of genotypic and environmental response to androgenesis in *N. rustica, Theor. Appl. Genet.,* 70, 128, 1985.

29. **Chu, C. C.,** The N₆ medium and its application to anther culture of cereal crops, in *Proc. Symp. Plant Tissue Culture,* Science Press, Peking, 1978, 43.

30. **Chu, C. C.,** Haploids in plant improvement, in *Plant Improvement and Somatic Cell Genetics,* Vasil, I. K., Scowcroft, W. R., and Frey, K. L., eds., Academic Press, New York, 1982, 129.

31. **Chu, C. C., Sun, C. S., and Wang, C. C.,** Cytological investigations on androgenesis of *Triticum aestivum, Acta Bot. Sin.,* 20, 6, 1978.

32. **Chu, C. C., Wang, C. C., Sun, C. S., Hau, C., Chian, N. F., Yin, K. C., Chu, C. Y., and Bi, F. Y.,** Establishment of an efficient medium for anther cultures of rice through comparative experiments on the nitrogen sources *Sci. Sin.,* 18, 659, 1975.

33. **Chuang, C. C., Ouyang, T. W., Chia, H., Chou, S. M., and Ching, C. K.,** A set of potato media for wheat anther culture, in *Proc. Symp. Plant Tissue Culture,* Science Press, Peking, 1978, 51.

34. **Clapham, D.,** Haploid *Hordeum* plants from anthers in vitro, *Z. Pflanzenzuecht.,* 69, 142, 1973.

35. **Collins, G. B., Dunwell, J. M., and Sunderland, N.,** Irregular microspore formation in *Datura innoxia* and its relevance to anther culture, *Protoplasma,* 82, 365, 1974.

36. **Corduan, G.,** Regeneration of anther-derived plants of *Hyoscyamus niger, Planta,* 127, 27, 1975.

37. **Dale, P. J.,** Pollen dimorphism and anther culture in barley, *Planta,* 127, 213, 1975.

38. **Dale, P. J. and Humphary, M. W.,** Rep. No. 88, Welsch Plant Breeding Station, Aberystwyth, Wales, 1974.

39. **Debergh, P. and Nitsch, C.,** Premiers resultats sur la culture in vitro de grains de pollen isoles chez la Tomate, *C.R. Acad. Sci.,* 276D, 1281, 1973.

40. **De Paepe, R., Belton, D., and Gnangbe, F.,** Basis and extent of genetic variability among doubled haploid plants obtained by pollen culture in *Nicotiana sylvestris, Theor. Appl. Genet.,* 59, 177, 1981.

41. **DeVerna, J. W. and Collins, G. B.,** Maternal haploids of *Petunia axillaris* via culture of placenta attached ovules, *Theor. Appl. Genet.,* 69, 187, 1984.

42. **Dollmantel, H. J. and Reinert, J.,** Auxin levels, antiauxins and androgenic plantlet formation in isolated pollen cultures of *Nicotiana tabacum, Protoplasma,* 103, 155, 1980.

43. **Duncan, E. J. and Heberle, E.,** Effect of temperature shock on nuclear phenomena in microspores of *Nicotiana tabacum* and consequently on plantlet production, *Protoplasma,* 90, 173, 1976.

44. **Dunwell, J. M.,** A comparative study of environmental and developmental factors which influence embryo induction and growth in cultured anthers of *Nicotiana tabacum, Environ. Exp. Bot.,* 16, 109, 1976.

45. **Dunwell, J. M.,** Anther culture in *Nicotiana tabacum.* The role of culture vessel atmosphere in pollen embryo induction and growth, *J. Exp. Bot.,* 30, 419, 1979.

46. **Dunwell, J. M.,** Stimulation of pollen embryo induction in tobacco by pretreatment of excised anthers in a water saturated atmosphere, *Plant Sci. Lett.,* 21, 9, 1981.

47. **Dunwell, J. M., Cornish, M., and de Courcel, A. G. L.,** Influence of genotype, plant growth temperature and anther incubation temperature on microspore embryo production in *Brassica napus* ssp. *oleifera, J. Exp. Bot.,* 36, 679, 1985.

48. **Dunwell, J. M. and Sunderland, N.,** Anther culture of *Solanum tuberosum, Euphytica,* 22, 317, 1973.

49. **Dunwell, J. M. and Sunderland, N.,** Pollen ultrastructure in anther culture of *Nicotiana tabacum.* I. Early stages of culture, *J. Exp. Bot.,* 25, 352, 1974.

50. **Dunwell, J. M. and Sunderland, N.,** Pollen ultrastructure in anther culture of *Nicotiana tabacum.* II. Changes associated with embryogenesis, *J. Exp. Bot.,* 25, 363, 1974.

51. **Dunwell, J. M. and Sunderland, N.**, Pollen ultrastructure in anther cultures of *Datura innoxia*. I. Division of the presumptive vegetative cell, *J. Cell Sci.*, 22, 469, 1976.

52. **Dunwell, J. M. and Thurling, N.**, Role of sucrose in microspore embryo production in *Brassica napus*, *J. Exp. Bot.*, 36, 1478, 1985.

53. **Durr, A. and Fleck, J.**, Production of haploid plants of *Nicotiana langsdorffii*, *Plant Sci. Lett.*, 18, 75, 1980.

54. **Engvild, K. C.**, Plantlet ploidy and flower bud size in tobacco anther cultures, *Hereditas*, 76, 320, 1974.

55. **Engvild, K. C., Linde-Laursen, I., and Lundquist, A.**, Anther culture of *Datura innoxia*, flower bud stage and embryoid level of ploidy, *Hereditas*, 72, 331, 1972.

56. **Erickson, R. O.**, Cytological and growth correlation in the flower bud and anther of *Lilium longiflorum*, *Am. J. Bot.*, 35, 729, 1948.

57. **Fitch, M. M. and Moore, P. H.**, Production of haploid *Saccharum spontaneum* — comparison of media for cell incubation of panicle branches and for float culture of anthers, *J. Plant Physiol.*, 117, 169, 1984.

58. **Forche, E., Kibler, R., and Neuman, K. H.**, The influence of developmental stages of haploid and diploid callus cultures of *Datura innoxia* on shoot initiation, *Z. Pflanzenphysiol.*, 101, 257, 1981.

59. **Foroughi-Wehr, B., Wilson, H. M., Mix, G., and Gaul, H.**, Monohaploid plants from anthers of dihaploid genotypes of *Solanum tuberosum*, *Euphytica*, 26, 361, 1977.

60. **Fouletier, B.**, Conditions favorisant la neoformation de cals haploides a partir d'antheres de Riz cultivers in vitro, *C.R. Acad. Sci. Ser. D*, 278, 2917, 1974.

61. **Geier, T. and Kohlenbach, H. W.**, Entwicklung von Embryonen und embryogenen kallus aus Pollenkornern von *Datura meteloides* and *Datura innoxia*, *Protoplasma*, 78, 381, 1973.

62. **Genovesi, A. D. and Magill, C. W.**, Improved rate of callus and green plant production from rice anther culture following cold shock, *Crop. Sci.*, 19, 662, 1979.

63. **George, L. and Rao, P. S.**, In vitro induction of pollen embryos and plantlets in *Brassica juncea* through anther culture, *Plant Sci. Lett.*, 26, 111, 1982.

64. **Ghandimathi, H.**, Direct pollen culture of *Nicotiana tabacum*, in *Proc. 5th Int. Congr. Plant Tissue and Cell Culture*, Fujiwara, A., Ed., Maruzen, Tokyo, 1982, 527.

65. **Gresshoff, P. M. and Doy, C. H.**, Haploid *Arabidopsis thaliana* callus and plants from anther culture, *Aust. J. Biol. Sci.*, 25, 259, 1972.

66. **Gresshoff, P. M. and Doy, C. H.**, Development and differentiation of haploid *Lycopersicon esculentum* (Tomato), *Planta*, 107, 161, 1972.

67. **Gresshoff, P. M. and Doy, C. H.**, Derivation of a haploid cell line from *Vitis vinifera* and the importance of the stage of meiotic development of anthers for haploid culture of this and other genera, *Z. Pflanzenphysiol.*, 73, 132, 1974.

68. **Guha, S. and Maheshwari, S. C.**, In vitro production of embryos from anthers of *Datura*, *Nature (London)*, 204, 497, 1964.

69. **Guha, S. and Maheshwari, S. C.**, Cell division and differentiation of embryos in the pollen grains of *Datura* in vitro, *Nature (London)*, 212, 97, 1966.

70. **Guha-Mukerjee, S.**, Genotypic differences in the in vitro formation of embryoids from rice pollen, *J. Exp. Bot.*, 24, 139, 1973.

71. **Gupta, P. P.**, Microspore-derived haploid, diploid and triploid plants in *Petunia violacea*, *Plant Cell Rep.*, 2, 255, 1983.

72. **Haccius, B. and Bhandari, N. N.**, Zur Frage der "Befestigung" junger pollen Embryonen von *Nicotiana tabacum* an der Antheren-Wand, *Beitr. Biol. Pflanz.*, 51, 53, 1975.

73. **He, Ding-Gang and Ouyang, Jun-Wen.**, Callus and plantlet formation from cultured wheat anthers at different developmental stages, *Plant Sci. Lett.*, 33, 71, 1984.

74. **Heberle-Bors, E.**, In vitro haploid formation from pollen: a critical review, *Theor. Appl. Genet.*, 71, 361, 1985.

75. **Heberle-Bors, E.**, Interaction of activated charcoal and iron chelates in anther cultures of *Nicotiana* and *Atropa belladonna*, *Z. Pflanzenphysiol.*, 99, 339, 1980.

76. **Heberle-Bors, E.**, Induction of embryogenic pollen grains in situ and subsequent in vitro pollen embryogenesis in *Nicotiana tabacum* by treatments of pollen donor plants with feminizing agents, *Physiol. Plant.*, 59, 67, 1983.

77. **Heberle-Bors, E.**, Genotypic control of pollen plant formation in *Nicotiana tabacum*, *Theor. Appl. Genet.*, 68, 475, 1984.

78. **Heberle-Bors, E. and Reinert, J.**, Androgenesis of isolated pollen cultures of *Nicotiana tabacum*: dependence upon pollen development, *Protoplasma*, 99, 237, 1979.

79. **Henry, Y., DeBuyser, J., Guenegou, T., and Ory, C.**, Wheat microspore embryogenesis during in vitro anther culture, *Theor. Appl. Genet.*, 67, 439, 1984.

80. **Henry, Y. and DeBuyser, J.**, Effect of IB/IR translocation on anther culture ability in wheat, *Triticum aestivum*, *Plant Cell Rep.*, 4, 307, 1985.

81. **Hepler, P. K. and Palevitz, B. A.,** Microtubules and microfilaments, *Ann. Rev. Plant Physiol.,* 25, 309, 1974.

82. **Hermundstad, S. A. and Peloquin, S. J.,** Germplasm enhancement with potato haploids, *J. Hered.,* 76, 463, 1985.

83. **Hinchee, M. A. W., Cruz, A. D., and Maretzki, A.,** Developmental and biochemical characteristics of cold-treated anthers of *Saccharum spontaneum, Z. Pflanzenphysiol.,* 115, 271, 1984.

84. **Hinchee, M. A. W. and Fitch, M. M. M.,** Culture of isolated microspores of *Saccharum spontaneum, Z. Pflanzenphysiol.,* 113, 305, 1984.

85. **Ho, K. M. and Jones, G. E.,** Mingo barley, *Can. J. Plant Sci.,* 60, 279, 1980.

86. **Horner, M. and Street, H. E.,** Pollen dimorphism — origin and significance in pollen plant formation by anther culture, *Ann. Bot.,* 42, 763, 1978.

87. **Hougas, R. W., Peloquin, S. J., and Ross, R. W.,** Haploids of the common potato, *J. Hered.,* 47, 103, 1958.

88. **Hsu, H. L. and Hung, W. L.,** Influence of 2,4-D and NAA on anther culture in rice, in *Proc. Symp. Anther Culture,* Science Press, Peking, 1978, 267.

89. **Hu, H., Hsi, T.-Y., Tseng, C.-C., Ouyang, T.-W., and Ching, C.-K.,** Application of anther culture to crop plants, in *Frontiers of Plant Tissue Culture,* Thorpe, T. A., Ed., Int. Assoc. Plant Tissue Culture, Calgary, 1978, 123.

90. **Huang, B. and Sunderland, N.,** Temperature stress pretreatment in barley anther culture, *Ann. Bot.,* 49, 77, 1982.

91. **Hughes, K. W., Bell, S. L., and Caponetti, J. D.,** Anther-derived haploids of the African violet, *Can. J. Bot.,* 53, 1442, 1975.

92. **Hunter, C. P.,** The effect of anther orientation on the production of microspore-derived embryoids and plants of *Hordeum vulgare* cv. Sabarlis, *Plant Cell Rep.,* 4, 267, 1985.

93. **Imamura, J. and Harada, H.,** Effects of abscisic acid and water stress on the embryo and plantlet formation in anther culture of *Nicotiana tabacum* cv. Samsun, *Z. Pflanzenphysiol.,* 100, 285, 1980.

94. **Imamura, J. and Harada, H.,** Studies on the changes in the volume and proliferation rate of cells during embryogenesis of in vitro cultured pollen grains of *Nicotiana tabacum, Z. Pflanzenphysiol.,* 96, 261, 1980.

95. **Imamura, J. and Harada, H.,** Stimulation of tobacco pollen embryogenesis by anaerobic treatments, *Z. Pflanzenphysiol.,* 103, 259, 1981.

96. **Imamura, J., Okaba, E., Kyo, M., and Harada, H.,** Embryogenesis and plantlet formation through direct culture of isolated pollen of *Nicotiana tabacum* cv. Samsun and *N. rustica* cv. *Rustica, Plant Cell Physiol.,* 23, 713, 1982.

97. **Irikura, Y.,** Induction of haploid plants by anther culture in tuber-bearing species and intergeneric hybrids of *Solanum, Potato Res.,* 18, 133, 1975.

98. **Jacobson, E. and Sopory, S. K.,** The influence and possible recombination of genotypes on the production of microspore embryoids in anther cultures of *Solanum tuberosum* and dihaploid hybrids, *Theor. Appl. Genet.,* 52, 119, 1978.

99. **Jensen, C. J.,** Monoploid production by chromosome elimination, in *Applied and Fundamental Aspects of Plant Cell Tissue and Organ Culture,* Reinert, J. and Bajaj, Y. P. S., Eds., Springer-Verlag, Berlin, 1977, 299.

100. **Jensen, W. A.,** Reproduction in flowering plants, in *Dynamic Aspects of Plant Ultrastructure,* Robards, A. W., Ed., McGraw-Hill, London, 1974, 483.

101. **Johansson, L.,** Effects of activated charcoal in anther cultures, *Physiol. Plant.,* 59, 397, 1983.

102. **Johansson, L. and Eriksson, T.,** Induced embryo formation in anther cultures of several *Anemone* species, *Physiol. Plant.,* 40, 172, 1977.

103. **Johansson, L. and Anderson, B.,** Effects of carbon dioxide in anther cultures, *Physiol. Plant.,* 60, 26, 1984.

104. **Johansson, L., Andersson, B., and Eriksson, T.,** Improvement of anther culture technique: activated charcoal bound in agar medium in combination with liquid medium and elevated CO_2 concentration, *Physiol. Plant.,* 54, 24, 1982.

105. **Kameya, T. and Hinata, K.,** Induction of haploid plants from pollen grains of *Brassica, Jpn. J. Breed.,* 20, 82, 1970.

106. **Kao, K. N.,** Plant formation from barley anther cultures with Ficoll media, *Z. Pflanzenphysiol.,* 103, 437, 1981.

107. **Kasha, K. J.,** Haploids from somatic cells, in *Haploids in Higher Plants, Advances and Potentials,* Kasha, K. J., Ed., University of Guelph, Canada, 1974, 67.

108. **Kasha, K. J. and Kao, K. N.,** High frequency haploid production in barley *(Hordeum vulgare), Nature (London),* 225, 874, 1970.

109. **Kasha, K. J. and Reinbergs, E.,** Utilization of haploids in barley, in *Barley Genetics,* Gaul, H., Ed., Verlag Karl Thiemig, Munich, 1975, 307.

110. **Kasha, K. J. and Reinbergs, E.,** Recent developments in the production and utilization of haploids in barley, in *Barley Genetics,* IV, Proc. 4th Int. Barley Genetics Symp., Asher, M. J., Ed., Churchill Livingstone, Edinburgh, 1981, 655.

111. **Kasperbauer, M. J. and Collins, G. B.,** Reconstitution of dihaploids from leaf tissue of anther-derived haploids in tobacco, *Crop Sci.,* 12, 98, 1972.

112. **Kaul, B. L. and Zutschi, U.,** Dimethyl sulphoxide as an adjuvant of colchicine in the production of polyploids in crop plants, *Indian J. Exp. Biol.,* 9, 522, 1971.

113. **Keller, W. A. and Armstrong, K. C.,** High frequency production of microspore-derived plants from *Brassica napus* anther cultures, *Z. Pflanzenzuecht,* 80, 100, 1978.

114. **Keller, W. A. and Armstrong, K. C.,** Stimulation of embryogenesis and haploid production in *Brassica campestris* anther cultures by elevated temperature treatments, *Theor. Appl. Genet.,* 55, 65, 1979.

115. **Keller, W. A., Rajhathy, T., and Lacapra, J.,** In vitro production of plants from pollen in *Brassica campestris, Can J. Genet. Cytol.,* 17, 655, 1975.

116. **Keller, W. A. and Stringam, G. R.,** Production and utilization of microspore-derived haploid plants, in *Frontiers of Plant Tissue Culture,* Thorpe, T. A., Ed., Int. Assoc. Plant Tissue Culture, Calgary, 1978, 113.

117. **Kermicle, J. L.,** Androgenesis conditioned by a mutation in maize, *Science,* 166, 1422, 1969.

118. **Kimber, G. and Riley, R.,** Haploid angiosperms, *Bot. Rev.,* 90, 480, 1963.

119. **Kochhar, T. S., Bhalla, P. R., and Sabharwal, P. S.,** Production of homozygous diploid plants by tissue culture technique, *J. Hered.,* 62, 59, 1971.

120. **Kohler, F. and Wenzel, G.,** Regeneration of isolated barley microspores in conditioned media and trials to characterize the responsible factor, *J. Plant Physiol.,* 121, 181, 1985.

121. **Konar, R. N.,** A haploid tissue from the pollen of *Ephedra foliata, Phytomorphology,* 13, 170, 1963.

122. **Konar, R. N. and Singh, M. N.,** Production of plantlets from female gametophyte of *Ephedra foliata, Z. Pflanzenphysiol.,* 95, 87, 1979.

123. **Ku, M. K., Cheng, W. C., Kuo, L. C., Kuan, Y. L., An, H. P., and Huang, C. H.,** Induction factors and morpho-cytological characteristics of pollen-derived plants in maize *(Zea mays)* , in *Proc. Symp. Plant Tissue Culture,* Science Press, Peking, 1978, 35.

124. **Kyo, M. and Harada, H.,** Studies on conditions for cell division and embryogenesis in isolated pollen culture of *Nicotiana rustica, Plant Physiol.,* 79, 90, 1985.

125. **Lacadena, J. R.,** Spontaneous and induced parthenogenesis and androgenesis, in *Haploids in Higher Plants, Advances and Potentials,* Kasha, K. J., Ed., University of Guelph, 1974, 13.

126. **LaRue, C. D.,** Studies on growth and regeneration in gametophytes and sporophytes of gymnosperms, *Brookhaven Symp. Biol.,* 6, 187, 1954.

127. **Lazar, M. D., Baenziger, P. S., and Schaeffer, G. W.,** Combining abilities and heritability of callus formation and plantlet regeneration in wheat *Triticum aestivum* anther cultures, *Theor. Appl. Genet.,* 68, 131, 1984.

128. **Lazar, M. D., Schaeffer, G. W., and Baenziger, P. S.,** The physical environment in relation to high frequency callus and plantlet development in anther culture of wheat *Triticum aestivum* cv. Chris., *J. Plant Physiol.,* 121, 103, 1985.

129. **Leaver, C. J.,** Mitochondrial genes and male sterility in plants, *Trends Biochem. Sci.,* 5, 248, 1980.

130. **Leelavathi, S., Reddy, V. S., and Sen, S. K.,** Somatic cell genetics in *Brassica* species. I. High production of haploid plants in *Brassica alba, Plant Cell Rep.,* 3, 102, 1984.

131. **Levings, C. S.,** Cytoplasmic male sterility, in *Genetic Engineering of Plants,* Kosuge, T., Meredith, C., and Hollaender, A., Eds., Plenum Press, New York, 1983, 81.

132. **Lichter, R.,** Anther culture of *Brassica napus* in a liquid culture medium, *Z. Pflanzenphysiol.,* 103, 229, 1981.

133. **Lichter, R.,** Induction of haploid plants from isolated pollen of *Brassica napus, Z. Pflanzenphysiol.,* 105, 427, 1982.

134. **Linsmaier, F. M. and Skoog, F.,** Organic growth factor requirements of tobacco tissue cultures, *Physiol. Plant.,* 18, 100, 1965.

135. **MacDonald, I. M. and Grant, W. F.,** Anther culture of pollen containing etherel induced micronuclei, *Z. Pflanzenzuecht.,* 73, 292, 1974.

136. **Maggoon, M. L. and Khanna, K. R.,** Haploids, *Caryologia,* 16, 191, 1963.

137. **Maheshwari, S. C., Rashid, A., and Tyagi, A. K.,** Haploids from pollen grains — retrospect and prospect, *Am. J. Bot.,* 69, 865, 1982.

138. **Maheshwari, S. C., Tyagi, A. K., Malhotra, K., and Sopory, S. K.,** Induction of haploidy from pollen grains in angiosperms — the current status, *Theor. Appl. Genet.,* 58, 193, 1980.

139. **Martineau, B., Hanson, M. R., and Ausubel, F. M.,** Effect of charcoal and hormones on anther culture of *Petunia* and *Nicotiana, Z. Pflanzenphysiol.,* 102, 109, 1981.

140. **Mascarenhas, J. P.,** The biochemistry of angiosperm pollen development, *Bot. Rev.,* 41, 259, 1975.

141. **Miah, M. A. A., Earle, E. D., and Khush, G. S.,** Inheritance of callus formation ability in anther cultures of rice *Oryza sativa, Theor. Appl. Genet.,* 70, 113, 1985.

142. **Michellon, R., Hugard, J., and Jonard, R.,** Sur l'isolement de colonies tissulaires de pecher *(Prunus persica* cv. 'Dixired' et Nectared IV) et d'Amandier *(Prunus amygdalus* cv. 'Al') a partir d'antheres cultivees in vitro, *C.R. Acad. Sci. Ser. D,* 278, 1719, 1974.

143. **Mii, M.,** Relationships between anther browning and plantlet formation in anther cultures of *Nicotiana tabacum, Z. Pflanzenphysiol.,* 80, 206, 1976.

144. **Mii, M.,** Effect of pollen degeneration on the pollen embryogenesis in anther cultures of *Nicotiana tabacum, Z. Pflanzenphysiol.,* 99, 349, 1980.

145. **Mio, S. H., Kuo, C. S., Kwei, Y. L., Sun, A. T., Ku, S. Y., Lu, W. I., Wang, Y. Y., Chen, M. I., Wu, M. K., and Hang, L.,** Induction of pollen plants of maize and observations on their progeny, in *Proc. Symp. Plant Tissue Culture,* Science Press, Peking, 1978, 23.

146. **Mitchell, A. Z., Hanson, M. R., Skvirsky, R. C., and Ausubel, F. M.,** Anther culture of *Petunia,* genotypes with high frequency of callus, root and plantlet formation, *Z. Pflanzenphysiol.,* 100, 131, 1980.

147. **Mix, G., Wilson, H. M., and Foroughi-Wehr, B.,** The cytological status of plants of *Hordeum vulgare* regenerated from microspore callus, *Z. Pflanzenzuecht.,* 80, 89, 1978.

148. **Montelongo-Escobedo, H. and Rowe, P. R.,** Haploid induction in potato: cytological basis for the pollinator effect, *Euphytica,* 18, 116, 1969.

149. **Murashige, T. and Skoog, F.,** A revised medium for rapid growth and bio-assays with tobacco tissue cultures, *Physiol. Plant.,* 15, 473, 1962.

150. **Nair, S., Gupta, P. K., and Mascarenhas, A. F.,** Haploid plants from in vitro anther culture of *Annona squamosa, Plant Cell Rep.,* 2, 198, 1983.

151. **Nakamura, A. and Itagaki, R.,** Anther culture in *Nicotiana* and the characteristics of the haploid plants, *Jpn. J. Breed.,* 23, 71, 1973.

152. **Nakamura, A., Yamada, T., Kadotani, N., and Itagaki, R.,** Improvement of flue-cured tobacco variety MC1610 by means of haploid breeding method and some problems of the method, in *Haploids in Higher Plants; Advances and Potentials,* Kasha, K. J., Ed., University of Guelph, Canada, 1974, 227.

153. **Narayanaswamy, S. and Chandy, L. P.,** In vitro induction of haploid, diploid and triploid androgenic embryoids and plantlets in *Datura metel, Ann. Bot.,* 35, 535, 1971.

154. **Niizeki, M. and Grant, W. F.,** Callus, plantlet formation, and ploidy from cultured anthers of *Lotus* and *Nicotiana, Can. J. Bot.,* 49, 2041, 1971.

155. **Niizeki, H. and Ono, K.,** Rice plants obtained by anther culture, in *Les Cultures de Tissus de Plantes (Colloq. Int. C.N.R.S.)* , No. 193, 1971, 251.

156. **Nitsch, C.,** Pollen culture — a new technique for mass production of haploid and homozygous plants, in *Haploids in Higher Plants: Advances and Potential,* Kasha, K. J., Ed., University of Guelph, Canada, 1974, 123.

157. **Nitsch, C.,** La culture de pollen isole sur milieu synthetique, *C.R. Acad. Sci. Ser. D,* 278, 1031, 1974.

158. **Nitsch, C.,** Single cell culture of an haploid cell: the microspore, in *Genetic Manipulation with Plant Material,* Ledoux, L., Ed., Plenum Press, New York, 1975, 297.

159. **Nitsch, C.,** Culture of isolated microspores, in *Applied and Fundamental Aspects of Plant Cell Tissue and Organ Culture,* Reinert, J. and Bajaj, Y. P. S., Eds., Springer-Verlag, Berlin, 1977, 268.

160. **Nitsch, C.,** Production of isogenic lines. Basic technical aspects of androgenesis, in *Plant Tissue Culture: Methods and Applications in Agriculture,* Thorpe, T. A., Ed., Academic Press, New York, 1981, 241.

161. **Nitsch, C. and Norreel, B.,** Effet d'un choc thermique sur le pouvoir embryogene du pollen de *Datura innoxia* cultive dans l'anthere ou isole de L'anthere, *C.R. Acad. Sci. Ser. D,* 276, 303, 1973.

162. **Nitsch, J. P.,** Experimental androgenesis in *Nicotiana, Phytomorphology,* 19, 389, 1969.

163. **Nitsch, J. P. and Nitsch, C.,** Haploid plants from pollen grains, *Science,* 163, 85, 1969.

164. **Nitsche, W. and Wenzel, G.,** *Haploids in Plant Breeding,* P. Parey, Berlin, 1977.

165. **Norstog, K.,** Induction of apogamy in megagametophytes of *Zamia integrifolia, Am. J. Bot.,* 52, 993, 1965.

166. **Ono, K. and Harashima, S.,** Induction of haploid callus from isolated microspores of Peony in vitro, *Plant Cell Physiol.,* 22, 337, 1981.

167. **Ouyang, T. W., Hu, H., Chuang, C.-C., and Tseng, C.-C.,** Induction of pollen plants from anthers of *Triticum aestivum* cultured in vitro, *Sci. Sin.,* 16, 79, 1973.

168. **Pelletier, G.,** Les conditions et les premiers stades de L'androgenese in vitro chez *Nicotiana tabacum, Mem. Soc. Bot. Fr. Colloq. Morphol.,* 261, 1973.

169. **Pelletier, G. and Ilami, M.,** Les facteurs de l'androgénèsés in vitro chez *Nicotiana tabacum, Z. Pflanzenphysiol.,* 68, 97, 1972.

170. **Perez-Bermudez, P., Cornejo, M. J., and Segura, J.,** Pollen plant formation from anther cultures of *Digitalis obscura, Plant Cell Tissue Organ Culture,* 5, 63, 1985.

171. **Phillips, G. C. and Collins, G. B.,** The influence of genotype and environment on haploid plant production from anther cultures of *Nicotiana tabacum, Tobbaco Inst.,* 179, 69, 1977.

172. **Picard, E.,** Influence de modifications dans les correlations internes sur la devenir du gametophyte male de *Triticum aestivum* in situ et an culture in vitro, *C.R. Acad. Sci. Ser. D,* 277, 777, 1973.

173. **Picard, E. and de Buyser, J.,** Nouveaux resultats concernant la culture d'antheres in vitro de ble tendre *(Triticum aestivum)* effects d'un choc thermique et de la position de l'anthere dans l'epi, *C.R. Acad. Sci. Ser. D,* 281, 127, 1975.

174. **Picard, E. and de Buyser,** High production of embryoids in anther cultures of pollen-derived homozygous spring wheat, *Ann. Amelior. Plant,* 27, 483, 1977.

175. **Pickering, R. A.,** The influence of genotype and environment on chromosome elimination in crosses between *Hordeum vulgare* × *Hordeum bulbosum, Plant Sci. Lett.,* 34, 153, 1984.

176. **Pickering, R. A. and Morgan, P. W.,** The influence of temperature on chromosome elimination during embryo development in crosses involving *Hordeum spp.* wheat *Triticum aestivum* and rye *Secale cereale, Theor. Appl. Genet.,* 70, 199, 1985.

177. **Plessers, A. G.,** Haploids as a tool in flax breeding, *Cereal News,* 8, 3, 1963.

178. **Primo-Millo, E. and Sunderland, N.,** Effect of plant growth temperature on pollen embryogenesis in *Nicotiana knightiana, John Innes Annu. Rep.,* 66, 233, 1976.

179. **Raghavan, V.,** Induction of haploid plants from anther culture of henbane, *Z. Pflanzenphysiol.,* 76, 89, 1975.

180. **Raghavan, V.,** Role of generative cell in androgenesis in henbane, *Science,* 191, 388, 1976.

181. **Raghavan, V.,** Origin and development of pollen embryoids and pollen calluses in cultured anther segments of *Hyoscyamus niger* (Henbane), *Am. J. Bot.,* 65, 984, 1978.

182. **Raghavan, V.,** Embryogenic determination and ribonucleic acid synthesis in pollen grains of *Hyoscyamus niger* (henbane), *Am. J. Bot.,* 66, 36, 1979.

183. **Raghavan, V.,** An autoradiographic study of RNA synthesis during pollen embryogenesis in *Hyoscyamus niger* (henbane), *Am. J. Bot.,* 66, 784, 1979.

184. **Raghavan, V.,** Distribution of poly(A)-containing RNA during normal pollen development and during induced pollen embryogenesis in *Hyoscyamus niger, J. Cell Biol.,* 89, 593, 1981.

185. **Raghavan, V.,** Protein synthetic activity during normal pollen development and during induced pollen embryogenesis in *Hyoscyamus niger, Can. J. Bot.,* 62, 2493, 1984.

186. **Rajhathy, T.,** Haploid flax revisited, *Z. Pflanzenzuecht.,* 76, 1, 1976.

187. **Rashid, A.,** Induction of embryos in pollen cultures of *Nicotiana sylvestris, Physiol. Plant,* 56, 223, 1982.

188. **Rashid, A.,** Pollen dimorphism in relation to pollen plant formation — a mini Review, *Physiol. Plant.,* 58, 549, 1983.

189. **Rashid, A. and Reinert, J.,** Selection of embryogenic pollen from cold-treated buds of *Nicotiana tabacum* var. Badisher Burley and their development into embryos in cultures, *Protoplasma,* 105, 161, 1980.

190. **Rashid, A. and Reinert, J.,** Differentiation of embryogenic pollen in cold-treated buds of *Nicotiana tabacum* var. Badischer Burley and nutritional requirements of the isolated pollen to form embryos, *Protoplasma,* 106, 137, 1981.

191. **Rashid, A. and Reinert, J.,** High-frequency embryogenesis in *ab initio* pollen cultures of *Nicotiana tabacum, Naturwissenschaften,* 68, 378, 1981.

192. **Rashid, A. and Reinert, J.,** In vitro differentiation of embryogenic pollen, control by cold treatment and embryo formation in *ab initio* pollen cultures of *Nicotiana tabacum* var. Badischer Burley, *Protoplasma,* 109, 285, 1981.

193. **Rashid, A. and Reinert, J.,** Factors affecting high-frequency embryo formation in *ab initio* pollen cultures of *Nicotiana, Protoplasma,* 116, 155, 1983.

194. **Rashid, A. and Street, H. E.,** The development of haploid embryoids from anther cultures of *Atropa belladonna, Planta,* 113, 263, 1973.

195. **Rashid, A. and Street, H. E.,** Segmentation in microspores of *Nicotiana sylvestris* and *N. tabacum* which lead to embryoid formation in anther cultures, *Protoplasma,* 80, 323, 1974.

196. **Rashid, A. and Street, H. E.,** Growth, embryogenic potential and stability of haploid cell culture of *Atropa belladonna, Plant Sci. Lett.,* 2, 89, 1974.

197. **Rashid, A., Siddiqui, A. W., and Reinert, J.,** Ultrastructure of embryogenic pollen of *Nicotiana tabacum* var. Badischer Burley, *Protoplasma,* 107, 375, 1981.

198. **Rashid, A., Siddiqui, A. W., and Reinert, J.,** Subcellular aspects of origin and structure of pollen embryos of *Nicotiana, Protoplasma,* 113, 202, 1982.

199. **Reinbergs, E., Park, S. J., and Song, L. S. P.,** Early identification of superior barley crosses by the double haploid technique, *Z. Pflanzenzuecht.,* 76, 215, 1976.

200. **Reinert, J., Bajaj, Y. P. S., and Heberle, E.,** Induction of haploid tobacco plants from isolated pollen, *Protoplasma,* 84, 191, 1975.

201. **Reynolds, T. L.,** An ultrastructural and stereological analysis of pollen grains of *Hyoscyamus niger* during normal ontogeny and induced embryogenic development, *Am. J. Bot.,* 71, 490, 1984.

202. **Reynolds, T. L.,** Ultrastructure of anomalous pollen development in embryogenic anther cultures of *Hyoscyamus niger, Am. J. Bot.,* 72, 44, 1985.

203. **Riley, R.,** The status of haploid research, in *Haploids in Higher Plants: Advances and Potential,* Kasha, K. J., Ed., University of Guelph, Canada 1974, 3.

204. **Raquin, C.,** Utilization of different sugars as carbon source for in vitro cultures of *Petunia, Z. Pflanzenphysiol.,* 111, 453, 1983.

205. **Sangwan-Norreel, B. S.,** Androgenic stimulating factors in the anther and isolated pollen grain culture of *Datura innoxia, J. Exp. Bot.,* 28, 843, 1977.

206. **Sangwan-Norreel, B. S.,** Male gametophyte nuclear DNA content evolution during androgenetic induction in *Datura innoxia, Z. Pflanzenphysiol.,* 111, 47, 1983.

207. **Sangwan, R. S. and Camefort, H.,** Cold-treatment related structural modification in the embryogenic anthers of *Datura, Cytologia,* 49, 473, 1984.

208. **SanNoeum, L. H.,** In vitro induction of gynoegenesis in higher plants, in *Proc. Conf. Broad. Genet. Base Crops,* Zeven, A. C. and Van Harten, A. M., Eds., Pudoc, Wageningen, 1978, 327.

209. **Schaeffer, G. W.,** Recovery of heritable variability in anther derived doubled haploid rice, *Crop. Sci.,* 22, 1160, 1982.

210. **Shannon, P. R. M., Nicholson, A. E., Dunwell, J. M., and Davies, D. R.,** Effect of anther orientation on microspore callus production in barley *(Hordeum vulgare), Plant Cell Tissue Organ Culture,* 4, 271, 1985.

211. **Sharma, D. R. and Chowdhury, J. B.,** Effects of different media on cultured anthers of *Datura innoxia* Mill., and comparative morphogenetic potentiality of haploid and diploid tissues, *Indian J. Exp. Biol.,* 15, 616, 1977.

212. **Sharma, D. P., Firoozabady, E., Ayers, N. M., and Galbraith, D. W.,** Improvement of anther culture in *Nicotiana:* media, cultural conditions and flow cytometric determination of ploidy levels, *Z. Pflanzenphysiol.,* 111, 441, 1983.

213. **Sharp, W. R., Raskin, R. S., and Sommer, H. E.,** The use of Nurse culture in the development of haploid clones in tomato, *Planta,* 104, 357, 1972.

214. **Simpson, E., Snape, J. W., and Finch, R. A.,** Variation between *Hordeum bulbosum* genotypes in their ability to produce haploids of barley, *Hordeum vulgare, Z. Pflanzenzuecht.,* 85, 205, 1980.

215. **Sopory, S. K.,** Physiology of Development of Pollen Embryoids in *Datura innoxia,* Ph.D. thesis, University of Delhi, 1972.

216. **Sopory, S. K.,** Effect of sucrose, hormones and metabolic inhibitors on the development of pollen embryoids in anther culture of dihaploid *Solanum tuberosum, Can. J. Bot.,* 57, 2691, 1979.

217. **Sopory, S. K., Jacobson, E., and Wenzel, G.,** Production of monohaploid embryoids and plantlets in cultured anthers of *Solanum tuberosum, Plant Sci. Lett.,* 12, 47, 1978.

218. **Sopory, S. K. and Maheshwari, S. C.,** Similar effects of iron chelating agents and cytokinins in the production of haploid embryos from the pollen grains of *Datura innoxia, Z. Pflanzenphysiol.,* 69, 97, 1973.

219. **Sopory, S. K. and Maheshwari, S. C.,** Morphogenetic potentialities of haploid and diploid vegetative parts of *Datura innoxia, Z. Pflanzenphysiol.,* 77, 274, 1975.

220. **Sopory, S. K. and Maheshwari, S. C.,** Development of pollen embryoids in anther cultures of *Datura innoxia.* I. General observations and effects of physical factors, *J. Exp. Bot.,* 27, 49, 1976.

221. **Sopory, S. K. and Maheshwari, S. C.,** Development of pollen embryoids in anther cultures of *Datura innoxia.* II. Effects of growth hormones, *J. Exp. Bot.,* 27, 58, 1976.

222. **Stettler, R., Bawa, K., and Levingston, G.,** Experimental induction of haploid parthenogenesis in forest trees, in *Induced Mutation in Plants,* International Atomic Energy Agency, Vienna, 1969, 661.

223. **Subhashini, U. and Venkateswarlu, T.,** Genotypes and their response to in vitro production of haploids in F_1 lines of *Nicotiana tabacum, Theor. Appl. Genet.,* 70, 225, 1985.

224. **Subrahmanyam, N. C. and Kasha, K. J.,** Selective chromosome elimination during haploid formation in barley following interspecific hybridization, *Chromosoma,* 42, 111, 1973.

225. **Subrahmanyam, N. C., Kasha, K. J., and Ali, A.,** Effect of gibberellic acid treatment and nutrient supply through detached tillers upon haploid frequency in barley, *Theor. Appl. Genet.,* 51, 169, 1978.

226. **Sun, C. C., Chu, C. C., Wang, C. C., and Tigerstedt, P. M. A.,** Studies on the anther culture of Triticale, *Acta Bot. Sin.,* 22, 27, 1980.

227. **Sun, C. S., Wang, C. C., and Chu, C. C.,** The ultrastructure of plastids in the albino pollen plants of rice, *Sci. Sin.,* 17, 793, 1974.

228. **Sun, C. S., Wu, S. C., Wang, C. C., and Chu, C. C.,** The deficiency of soluble proteins and plastid ribosomal RNA in the albino pollen plantlets of rice, *Theor. Appl. Genet.,* 55, 193, 1979.

229. **Sunderland, N.,** Anther culture: a progress report, *Sci. Prog.,* 59, 527, 1971.

230. **Sunderland, N.,** Anther culture as a means of haploid induction, in *Haploids in Higher Plants: Advances and Potential,* Kasha, K. J., Ed., University of Guelph, Canada, 1974, 91.

231. **Sunderland, N.,** Strategies in the improvement of yields in anther culture, in *Proc. Symp. Plant Tissue Culture,* Science Press, Peking, 1978, 65.

232. **Sunderland, N.,** Comparative studies of anther and pollen culture, in *Plant Cell and Tissue Culture: Principles and Applications,* Sharp, W. R., Larsen, P. O., Paddock, E. F., and Raghavan, V., Eds., Ohio State University Press, Columbus, 1979, 203.

233. **Sunderland, N.,** Induction of growth in the culture of pollen, in *Differentiation in Vitro,* Yeoman, M. M. and Truman, D. E. S., Eds., Cambridge University Press, 1982, 1.

234. **Sunderland, N.,** On the use of microspores for genetic modification, in *Genetic Engineering of Plants,* Kosuge, T., Meredith, C., and Hollaender, A., Eds., Plenum Press, New York, 1983, 315.

235. **Sunderland, N., Collins, G. B., and Dunwell, J. M.,** The role of nuclear fusion in pollen embryogenesis of *Datura innoxia, Planta,* 117, 227, 1974.

236. **Sunderland, N. and Dunwell, J. M.,** Anther and pollen culture in *Plant Tissue and Cell Culture,* Street, H. E., Ed., Blackwell, Oxford, 1977, 223.

237. **Sunderland, N. and Evans, L. I.,** Multicellular pollen formation in cultured barley anthers. II. The A, B and C pathways, *J. Exp. Bot.,* 31, 501, 1980.

238. **Sunderland, N., Huang, B., and Hill, G. J.,** Disposition of pollen in situ and its relevance to anther/pollen culture, *J. Exp. Bot.,* 35, 521, 1984.

239. **Sunderland, N. and Roberts, M.,** New approach to pollen culture, *Nature (London),* 270, 236, 1977.

240. **Sunderland, N. and Roberts, M.,** Cold pre-treatment of excised flower buds in float cultures of tobacco anthers, *Ann. Bot.,* 43, 405, 1979.

241. **Sunderland, N. and Wicks, F. M.,** Embryoid formation in pollen grains of *Nicotiana tabacum, J. Exp. Bot.,* 22, 213, 1971.

242. **Sunderland, N. and Wildon, D. C.,** A note on the pretreatment of excised flower buds in float cultures of *Hyoscyamus* anthers, *Plant Sci. Lett.,* 15, 1979.

243. **Symko, S.,** Haploid barley from crosses of *Hordeum bulbosum* $(2 \times)$ $\times$ *H. vulgare* $(2 \times)$, *Can J. Genet. Cytol.,* 11, 602, 1969.

244. **Tan, B. H. and Halloran, G. M.,** Pollen dimorphism and the frequency of inductive anthers in anther cultures of *Triticum monococcum, Biochem. Physiol. Pflanz.,* 177, 197, 1982.

245. **Thompson, K. F.,** Frequencies of haploids in spring oil-seed rape (*Brassica napus*), *Heredity, 24, 318, 1969.*

246. **Thompson, K. F.,** Haploid breeding technique for flax, *Crop. Sci.,* 17, 757, 1977.

247. **Thurling, N. and Chay, P. M.,** The influence of donor plant genotype and environment on production of multicellular microspores in cultured anthers of *Brassica napus* ssp. Oleifera, *Ann. Bot.,* 54, 681, 1984.

248. **Ting, Y. C., Yu, M., and Zheng, W.-Z.,** Improved anther culture of maize (*Zea mays*), *Plant Sci. Lett.,* 23, 139, 1981.

249. **Tomes, D. T. and Collins, G. B.,** Factors affecting haploid plant production from in vitro anther cultures of *Nicotiana* species, *Crop. Sci.,* 16, 837, 1976.

250. **Torrizo, L. B. and Zapata, F. J.,** Anther culture in rice. IV. The effect of abscisic acid on plant regeneration, *Plant Cell Rep.,* 5, 136, 1986.

251. **Tran Thanh Van, K. and Hanh, T. T.,** Embryogenic capacity of anthers from flowers formed in vitro on thin cell layers and of anthers excised from mother plant of *Nicotiana tabacum* and *N. plumbaginifolia, Z. Pflanzenphysiol.,* 100, 379, 1980.

252. **Tsay, S. S., Tsay, H. S., and Chao, C. Y.,** Cytochemical studies of callus development from microspores in cultured anther of rice, *Plant Cell Rep.,* 5, 119, 1986.

253. **Tsun, C. Y.,** in *Proc. Sym. Anther Culture,* Science Press, Peking, 1978, 297.

254. **Tsunewaki, K., Noda, K., and Fujisawa, T.,** Haploid and twin formation in a wheat strain Salmon with alien cytoplasm, *Cytologia,* 33, 526, 1968.

255. **Tulecke, W.,** A tissue derived from the pollen of *Ginkgo biloba, Science,* 117, 599, 1953.

256. **Tulecke, W.,** The pollen cultures of C. D. LaRue: a tissue from the pollen of *Taxus, Bull. Torrey Bot. Club,* 86, 283, 1959.

257. **Tulecke, W. and Sehgal, N.,** *Contrib. Boyce Thompson Inst.,* 22, 1963.

258. **Turcotta, E. L. and Feaster, C. V.,** Semigametic production of cotton haploids, in *Haploids in Higher Plants: Advances and Potentials,* Kasha, K. J., Ed., University of Guelph, Canada, 1974, 53.

259. **Tyagi, A. K., Rashid, A., and Maheshwari, S. C.,** High-frequency production of embryos in *Datura innoxia* from isolated pollen grains by combined cold treatment and serial culture of anthers in liquid medium, *Protoplasma,* 99, 11, 1979.

260. **Tyagi, A. K., Rashid, A., and Maheshwari, S. C.,** Enhancement of pollen embryo formation in *Datura innoxia* by charcoal, *Physiol. Plant.,* 49, 296, 1980.

261. **Tyagi, A. K., Rashid, A., and Maheshwari, S. C.,** Promotive effect of polyvinylpolypyrrolidone on Pollen embryogenesis in *Datura innoxia, Physiol. Plant.,* 53, 405, 1981.

262. **Vasil, I. K.,** Physiology and cytology of anther development, *Biol. Rev.,* 42, 327, 1967.

263. **Vasil, I. K.,** Androgenic haploids. *Perspectives in plant cell and tissue culture, Int. Rev. Cytol.,* Suppl. 11A, 195, 1980.

264. **Wang, P. and Chen, Y. R.,** Effects of growth conditions on anther donor plants on the production of pollen plants in wheat anther culture, *Acta Genet. Sin.,* 7, 64, 1980.

265. **Wang, C. C., Sun, C. S., and Chu, Z. C.,** On the conditions for the induction of rice pollen plants and certain factors affecting the frequency of induction, *Acta Bot. Sin.,* 16, 43, 1974.

266. **Wang, C. C., Sun, C. S., and Chu, C. C.,** Effect of culture factors in vitro on the production of albino pollen plants in rice, *Acta Bot. Sin.,* 19, 190, 1977.

267. **Wang, C. C., Sun, C. S., Chu, S. C., and Wu, S. C.,** Studies on the albino pollen plantlets of rice, in *Proc. Symp. Plant Tissue Culture,* Science Press, Peking, 1978, 149.

268. **Weatherhead, M. A. and Henshaw, G. G.,** The induction of embryoids in free pollen culture of potatoes, *Z. Pflanzenphysiol.,* 94, 441, 1979.

269. **Weatherhead, M. A., Burdon, J., and Henshaw, G. G.,** Some effects of activated charcoal as an additive to plant tissue culture media, *Z. Pflanzenphysiol.,* 89, 141, 1978.

270. **Wei, Z. M., Kyo, M., and Harada, H.,** Callus formation and plant regeneration through direct culture of isolated pollen of *Hordeum vulgare* cv. Sabarlis, *Theor. Appl. Genet.,* 72, 252, 1986.

271. **Wenzel, G., Hoffman, F., and Thomas, E.,** Increased induction and chromosome doubling of androgenetic haploid rye, *Theor. Appl. Genet.,* 51, 81, 1977.

272. **Wenzel, G., Hoffman, F., Potrykus, I., and Thomas, E.,** The separation of viable rye microspores from mixed populations and their development in culture, *Mol. Gen. Genet.,* 138, 293, 1975.

273. **Wenzel, G. and Thomas, E.,** Observations on the growth in cultures of anthers of *Secale cereale, Z. Pflanzenzuecht.,* 72, 89, 1974.

274. **Wernicke, W. and Kohlenbach, H. W.,** Investigations on liquid culture medium as a means of anther culture in *Nicotiana, Z. Pflanzenphysiol.,* 79, 189, 1976.

275. **Wernicke, W. and Kohlenbach, H. W.,** Experiments on the culture of isolated microspores in *Nicotiana* and *Hyoscyamus, Z. Pflanzenphysiol.,* 81, 330, 1977.

276. **Wernicke, W., Harms, C. T., Lorz, H., and Thomas, E.,** Selective enrichment of embryogenic microspore populations, *Naturwissenschaften,* 65, 540, 1978.

277. **White, P. R.,** *The Cultivation of Animal and Plant Cells,* Ronald Press, New York, 1963.

278. **Wilson, H. M., Mix, G., and Foroughi-Wehr, B.,** Early microspore divisions and subsequent formation of microspore calluses at high frequency in anthers of *Hordeum vulgare, J. Exp. Bot.,* 29, 227, 1978.

279. **Wu, J. L., Zhung, Q. L., Nong, F. R., and Chang, T. M.,** *Hereditas (Bejing),* 2, 23, 1980.

280. **Xu, Z. H., Huang, B., and Sunderland, N.,** Culture of barley anthers in conditioned media, *J. Exp. Bot.,* 32, 767, 1981.

281. **Xu, Z. H. and Sunderland, N.,** Glutamine, inositol and conditioning factor in the production of barley pollen callus in vitro, *Plant Sci. Lett.,* 23, 161, 1981.

282. **Yin, K. C., Hsu, C., Chu, C. Y., Bi, Y., Wang, S. T., Liu, T. Y., Chu, C. C., Wang, C. C., and Sun, C. S.,** A study of new cultivar of rice raised by haploid breeding method, *Sci. Sin.,* 19, 227, 1976.

283. **Zamir, D., Jones, R. A., and Kedar, N.,** Anther culture of male sterile tomato *(Lycopersicon esculentum)* mutants, *Plant Sci. Lett.,* 17, 353, 1980.

284. **Zenkteler, M.,** In vitro production of haploid plants from pollen grains of *Atropa belladonna, Experientia,* 27, 1087, 1971.

285. **Zhou, J. Y.,** Pollen dimorphism and its relation to formation of pollen embryos in anther culture of wheat *Triticum aestivum, Acta Bot. Sin.,* 22, 117, 1980.

286. **Zhou, C. and Yang, H.-Y.,** Induction of haploid rice plantlets by ovary culture, *Plant Sci. Lett.,* 20, 231, 1981.

287. **Zhu, Z. C. and Wu, H. S.,** In vitro induction of haploid plantlets from unpollinated ovaries of *Triticum aestivum* and *Nicotiana tabacum, Acta Genet. Sin.,* 6, 181, 1979.

288. **Zhuang, J., Ku, J., Chen, G., and Sun, S.,** Factors affecting the induction of pollen plants of intergeneric hybrids of *Triticum aestivum* × *Triticum* Agropyron, *Theor. Appl. Genet.,* 70, 294, 1985.

INDEX

Printed and bound by CPI Group (UK) Ltd, Croydon, CR0 4YY

22/10/2024

01777633-0019